Law of Health and Safety at Work

Norman Selwyn

JP, LLM, Dip Econ (Oxon), ACIS, Barrister
Lecturer in Law at the University of Aston
in Birmingham

London
BUTTERWORTHS
1982

England London Butterworth & Co (Publishers) Ltd
88 Kingsway, WC2B 6AB

Australia Sydney Butterworths Pty Ltd
271–273 Lane Cove Road, North Ryde, NSW 2113
Also at Melbourne, Brisbane, Adelaide and Perth

Canada Toronto Butterworth & Co (Canada) Ltd
2265 Midland Avenue, Scarborough, M1P 4S1

New Zealand Wellington Butterworths of New Zealand Ltd
33–35 Cumberland Place

South Africa Durban Butterworth & Co (South Africa) (Pty) Ltd
152–154 Gale Street

USA Boston Butterworth (Publishers) Inc
10 Tower Office Park, Woburn, Mass. 01801

ISBN 0 406 66750 0

Typeset and printed by Singapore National Printers Pte Ltd

Preface

Although the subject of health and safety at work is of vital importance to every employer, employee and self-employed person in this country, there is an apparent dearth of books which explain the complex legal requirements in a manner which can be readily appreciated by those who are most closely affected. The aim of this book is to fill that gap. It should be of interest to employers, company secretaries, managers, trade unionists, safety officers, safety representatives, enforcement officers and lawyers, as well as to students generally who will be seeking to enter employment at some future time.

The book is intended as a guide, not a bible. It seeks to state the law as it is generally applicable, but in all cases where legal problems arise in practice, the reader may need to consult the actual statutory provisions or, where necessary, take expert legal advice.

The *raison d'être* of health and safety legislation is not always easy to discover, for there are complex social objectives to be achieved. The law is intended to be partly preventative (evidenced by the power to issue various types of enforcement notices), partly compensatory (by means of a civil action for damages in respect of injuries received) and partly punitive (through the use of criminal sanctions). The rules are an amalgam of contract law, tort and crime. Statutory provisions are interwoven with judicial decisions, and in recent years Approved Codes of Practice have assumed an even greater importance in giving practical guidance. Each industry, each concern, has its own peculiar problems, and each incident has unique features which makes accident prevention a difficult task for those involved. But however complex these matters may be, ignorance of the legal requirements is the least excuse.

As this book is of a general nature, certain topics (such as mines and quarries legislation) have of necessity had to be omitted. Environmental issues also appear to be outside the self-imposed boundaries, though this does not imply that such matters are of lesser importance. I can only hope that my experience in lecturing to innumerable groups from all sides of industry and commerce has placed me on the right track so far as the actual contents are concerned.

By the provisions of the Social Security and Housing Benefits Bill, recently introduced into Parliament, industrial injuries benefits (see chapter 9) will be abolished, and a new employers' statutory sick pay scheme will come into force in April 1983.

NORMAN SELWYN
Management Centre
University of Aston
Birmingham
July 1981

Contents

Table of statutes

References to *Statutes* are to Halsbury's Statutes of England (Third Edition) showing the volume and page at which the annotated text of the Act will be found. References in the right-hand column are to paragraph numbers.

Table of cases

1 Law and legal institutions

The background to health and safety law

1.1 Legislative intervention in pursuance of the cause of health and safety dates from the Health and Morals of Apprentices Act 1802, which was designed to protect young children working in cotton and woollen mills, and other factories where more than twenty persons were employed. At that time it was the custom to put four children in a bed during the day, and four more in the same bed at night, while the day-time occupants were working, and it was this type of abuse at which the Act was aimed. Other pieces of minor legislation were passed in the ensuing years, but the real breakthrough came in 1833, when the Factory Act was passed. It provided that four factory inspectors were to be appointed, with powers of investigation and prosecution. From that time on, the factory movement gathered momentum, although the motives of some of the protagonists were not always of high altruism. Sometimes the legislation was designed to restrict the hours of work of women and young children (indirectly benefiting adult male labour), while other enactments were concerned with establishing safe working conditions. Perhaps the worst feature was the multiplicity of legislation – in 1876 a Factory Commission reported that the law was in a complete state of chaos, with no less than nineteen different enactments to be considered. It was not until the Factory and Workshop Act of 1901 that a comprehensive piece of factory legislation was enacted, when all the previous law was consolidated into one statute. One of the more interesting innovations was to give the power to the Secretary of State to make Regulations for particular industries, a power which was used extensively to control a large number of different industrial processes. In 1916 this power was extended to permit Welfare Orders to be made, dealing with washing facilities, first aid provisions, and so on. A further major breakthrough came in 1937 when the Factories Act was passed. It swept away the old distinctions between the different types of premises (textile factories, workshops, etc) and made detailed provisions for health, safety and welfare. Minor amendments were made in 1948 and 1959, and the various statutes were consolidated once more by the Factories Act 1961, which is still in force today.

1.2 The mining industry was an obvious subject for protective legislation, and in 1842 a Mines and Collieries Act was passed, which was mainly concerned with regulating hours and working conditions of women and children. Work in quarries was brought within the scope of the law by the Metalliferous Mines Regulations Act 1872, and a further major reform took place in 1911, with the passing of the Coal Mines Act. These provisions were strengthened and brought up to date in the Mines and Quarries Act 1954 which, together with Regulations made, lays down a comprehensive protective code of legislation for these industries.

1.3 Health and safety legislation for the non-industrial worker came about somewhat later. In 1886, the Shop Hours Regulation Act restricted the hours of work of young persons, and the Seats for Shop Assistants Act 1899 may be regarded as an important welfare statute. After many attempts had been made to give legislative cover to office workers, an Offices Act was passed in 1960, but this (and other legislation) was superseded by the Offices, Shops and Railway Premises Act 1963 (for which see chapter 5). Other statutes which regulated hours of work include the Employment of Women, Young Persons and Children Act 1920, Hours of Employment (Conventions) Act 1936, and Young Persons (Employment) Act 1938.

1.4 The original intention of the framers of the legislation was that health and safety laws would be enforced through the use of criminal sanctions, but for two reasons this emphasis changed over the years. The first was the gradual realisation that a rigorous policy of enforcement by the factory inspectorate would probably lead to the shut-down of large sections of British industry, already fighting hard to maintain its position against competition from foreign countries which were not necessarily inhibited by such constraints. The policy would also clutter up the courts and require a massive expansion of the machinery for bringing prosecutions. The second, and possibly more significant reason for the change in emphasis was the ability of injured workmen to bring civil actions

658-00

Radio Rentals

35-00.

100.00 weft wk
hkeep.

161.00..

40 B.T.
30 FA.
30 Sects.
45. Elec
50 Gas

165-00.

25 Sears
.00

200-00
317-00

517-00 -

247-00 Mgage
10-00 Ins.

260.00.

60.00 Sears & RR

200.00.

/////

Wednesday – must put 80·00 in Bank.

(100·00 if can). P.O.

will need to cover

100-00 (loan). ⇒ by Tuesday

cash to last.

<u>or</u> put less money 100·00 in Bank tomorrow

leaving ~~eat~~ enough to last till Friday. Mikes-money

⇒. If · can · ⇒ .

<u>Next week:</u> To come off: 180·00 cheques

(32101.05)

2851-50

AUG REC. (3427.18)

(575.68)

367-50

School cloth grants.

4122-22.

in respect of injuries suffered as a result of their employers' failure to observe the statutory duties. This action – for breach of statutory duty – was first successful in the case of *Groves v Lord Wimborne* in 1898, and, because of the absolute nature of many of the statutory duties, became a fruitful source of financial solace. But the result was to divert the objects of the law away from prosecution and prevention towards a system of civil compensation. As Goddard LJ stated in *Hutchinson v London and North Eastern Rly Co*, 'The real incentive for the observance by employers of their statutory duties. . . is not their liability to substantial fines, but the possibility of heavy claims for damages'. Even this, however, was not entirely correct, for most employers realised the value of taking out insurance policies against such claims (a requirement which was made compulsory in 1969) with the result that both the civil and criminal law ceased to be major deterrents.

1.5 At the same time as these developments were taking place, the common law of the country was also becoming active. In a leading case decided in 1837 (*Priestley v Fowler*) a claim based on an allegation of negligence by an employer towards his employee failed, but a few years later, in 1840, a young girl successfully sued when she was seriously injured in a mill accident, and was awarded damages (*Cottrell v Stocks*). From then on, common law claims were bogged down with problems arising from the doctrine of 'common employment' (abolished in 1948 by the Law Reform (Personal Injuries) Act), the inability to sue at common law if the injured workman had also claimed under the Workman's Compensation Act (amended by the National Insurance (Industrial Injuries) Act 1946), and the inability to claim if the employee had been contributorily negligent (amended by the Law Reform (Contributory Negligence) Act 1945). By 1970, the law on health and safety was largely concerned with issues of compensation.

1.6 In 1970 a Committee on Health and Safety at Work was appointed under the chairmanship of Lord Robens, which reported two years later. In their report, the Robens Committee reached some fundamental conclusions. First, although it was generally recognised that we had in this country the finest regulatory system of legal controls anywhere in the world, this did not prevent the annual carnage which took place each year, evidenced by the numbers who were killed or injured. Second, much of our law was obscure and unintelligible to those whose actions it was intended to influence. There was a haphazard mass, intricate in detail, difficult to amend, and frequently out-of-date. Third, the various enforcement authorities had overlapping jurisdictions which caused some confusion.

1.7 But the main conclusion of the Robens Committee was that there was one single cause, above all else, for accidents and ill health at work. This was, apathy. Apathy at the top, apathy at the bottom, apathy at all levels in between. True, there were a few dedicated people who worked hard at trying to influence people's attitudes, but these rarely had sufficient power or authority to override considerations of production, and in the main, health and safety had a low priority in almost all workplaces. In an attempt to overcome this attitude, the Robens Committee made a number of far-reaching proposals. The first was to devise a system whereby all emloyers and all employees became aware that health and safety was the concern of everyone, and not just a matter for the dedicated few. As the Report stated, 'Our present system encourages too much reliance on State regulation, and rather too little on personal responsibility and voluntary, self-generating effort.' Next, it was suggested that there was a need for a single comprehensive framework of legislation which would cover all work activity, supported and supplemented by a series of controls to deal with specific problems, and assisted by voluntary standards and more flexible Codes of Practice. Finally, a more unified enforcement authority was needed, having overall responsibility for initiating legal proposals and giving assistance and advice, possessing stronger enforcement powers, and with the ability to delegate its enforcement functions when necessary.

1.8 The result was the passing of the Health and Safety at Work etc Act 1974 (HSWA), an important piece of legislation which adopts a fundamentally different approach to the whole subject. In the first place, the Act applies to all employed persons, wherever they work. This has brought into the protective umbrella an estimated 8,000,000 'new entrants' who were not hitherto covered by previous legislation. Next, as well as laying down duties for employers and employees, the Act imposes certain legal requirements on those who manufacture, import, design or supply articles and substances which are to be used at work. Some of the provisions (e g safety policies, safety representatives) are designed to bring about a greater personal involvement of all concerned. New institutions were created, new enforcement powers were enacted, and new concepts, such as the use of Approved Codes of Practice, were introduced. The 'old' law, e g Factories Act, Offices, Shops and Railway Premises Act, etc, will still remain in force, although the ultimate objective is to gradually repeal and replace them by new

Regulations which will be supplemented by Approved Codes of Practice.

1.9 It must not be assumed that the Health and Safety at Work Act is the answer to all our problems, or that it is a perfect piece of legislation. On the contrary, despite the criticism levelled by the Robens Committee against the legalistic and unintelligible nature of the then existing law, the new Act is turgid, soporific, and, in parts, about as meaningful as medieval metaphysics. Further, it is basically a criminal statute, with appropriate penalties for breaches, and does not confer on anyone the right to make a claim for compensation. This means that most of the cases will start and finish in the magistrates' courts, with the result that there is very little opportunity for authoritative interpretations of the complex legal language. Those whose task it is to implement health and safety policies frequently yearn in vain for guidance.

1.10 On the positive side, the Act has acted as a catalyst for considerable management activity, as a greater awareness of responsibility has brought about increased concern for the health and safety of employees. It is, perhaps, too early to judge the psychological impact of the Act, or to evaluate its success, for although it is possible to note some downward trends in the numbers of accidents, there are many other factors which need to be taken into account, and we can never know how many accidents did not happen as a result of a more safety-conscious approach. Meanwhile, if the propaganda which surrounds the Act brings about a greater awareness that health and safety is everyone's concern, and not merely a matter for those who have a specialised interest in the problem, nothing but good will be the result.

1.11 The present situation, therefore, is that the law on health and safety at work is an amalgam of criminal law, civil law, and preventative measures. Breaches of the various statutory provisions (and Regulations made thereunder) are capable of being criminal offences, in respect of which the wrongdoer may be fined or (in rare circumstances) imprisoned. Civil claims for compensation may arise in respect of injuries received as a result of a failure by the employer to observe the statutory requirements (other than HSWA), or a failure to carry out those duties imposed by the common law of the land. The preventative measures can be found in the new enforcement notices which may be issued, and the greater involvement of the workforce by the appointment of safety representatives and the creation of safety committees.

1.12 This mixture of legal objectives is reflected in the decided cases which explain and expand on the nature of the legal duties. For practical reasons, the majority of these cases stem from civil claims, although in recent years there has been an increasing number of rulings in the industrial tribunals which add to our understanding of the legal rules. Care must always be taken not to elevate all such decisions to absolute principles, for the statute is always paramount, and individual cases can only be decided on their own special facts. Legal interpretation is more of an art than a science.

Sources of law

A LEGISLATION

1.13 The prime source of law in the United Kingdom consists of Acts of Parliament. With the exception noted below relating to the Treaty of Rome, an Act of Parliament is the supreme law of the land. Parliament, it is said, can make or unmake any law whatsoever. In strict theory, this could lead to severe problems, but in reality there are clearly political and practical limits on what Parliament in fact will do.

1.14 Statute law commences with the introduction of a Bill into either House of Parliament. This will receive a formal First Reading, when it is published. The House will then hold a full debate on the general principles behind the proposed legislation, and a vote may be held on a proposal to give the Bill a Second Reading. If successful, the Bill will then be sent to a Committee, where it is considered in detail, and amendments may be made. Next, the Bill is presented to a Report Stage of the full House, when the amendments are considered, and finally, the Bill will receive its Third Reading. It will then go to the other House, where a similar procedure is adopted. When both Houses have passed the Bill, it is presented to the Queen for the Royal Assent, which, by convention, is never refused.

1.15 However, although the Bill is now an Act, its implementation may be delayed in whole or in part. Modern legislation frequently contains powers which enable the appropriate Minister to bring the Act into force in various stages, often with transitional provisions. In those cases the law will not come into force until the date specified in a Commencement Order.

B REGULATIONS

1.16 An Act may confer upon a Minister the power to make new or additional law by means of Regulations. Such subordinate legislative power has many advantages, for it enables technical proposals to be passed, it is speedier than the procedure adopted for an Act, and enables the law to be amended by a simpler procedure. Regulations must be laid before Parliament, and are considered by the Joint Committee on Statutory Instruments, which has members from both Houses. They are not concerned with the policy or merits of the Regulations, only the technical competence of the Minister to make them. For example, the Committee will ascertain whether the Regulations are within the powers which are conferred by the parent Act, and whether they contain no unusual or unexpected use of that power. Some Regulations will only become law after an affirmative vote by both Houses of Parliament, but the majority will become law when they are made. However, they can subsequently be vetoed by a negative vote within forty days of laying. Regulations which are made under HSWA are of this type (see section 82(3)).

1.17 Regulations have the full force of law until they are repealed or amended in some way. It is possible for them to supersede specific provisions of the parent Act. In *Miller v William Boothman*, the plaintiff was injured while working on a circular saw, the fencing of which complied with the Woodworking Machinery Regulations 1922. It was argued on his behalf that the employers were still liable for his injury, as they were in breach of the requirements of section 14 of the Factories Act to fence every dangerous part of any machinery. The argument was rejected. The Minister had exercised the power vested in him to exclude the provisions of the Act by Regulation, and the latter prevailed.

1.18 As has already been indicated, one of the main objects of HSWA is to replace the existing statutory provisions over a period of time with a more streamlined system in which Regulations will play a major part. These will fall generally into three categories. First, there will be those Regulations which will apply to most or all employment situations (for example, the Notification of Accidents and Dangerous Occurrences Regulations 1980); second, there will be those Regulations which are designed to control a particular hazard in a particular industry (e g Agriculture (Threshers and Balers) Regulations 1960); third, there are those which will refer to a particular hazard or risk which may be found in a number of industries or processes (for example, the Control of Lead Regulations 1980). Sometimes, Regulations may be introduced in order to further the streamlining process or because the old provisions were out of date (e g the First Aid Regulations 1981), others may be introduced consequent upon the discovery of a loophole in the law which may have come to light because of some litigation (the Abrasive Wheels Regulations 1970 were made in order to nullify the decision of *Summers & Sons v Frost*). As a general rule, Regulations are designed to supplement and strengthen the provisions of the various Acts of Parliament, spelling out the requirements in greater detail (e g Power Press Regulations 1965).

1.19 Health and Safety Regulations made under HSWA, section 15 may be for any of the following purposes:

(a) repeal or modify any existing statutory provision;
(b) exclude or modify in relation to any specific class of case any of the provisions of sections 2–9 (chapter 3) or any existing statutory provision;
(c) make a specific authority responsible for the enforcement of any relevant statutory provision;
(d) impose requirements by reference to the approval of the Commission or other specified body or person;
(e) provide that any reference in a Regulation to a specific document shall include a reference to a revised version of that document;
(f) provide for exemptions from any requirement or prohibition;
(g) enable exemptions to be granted by a specified person or authority;
(h) specify the persons or class of persons who may be guilty of an offence;
(i) provide for specified defences either generally or in specified circumstances;
(j) exclude proceedings on indictment in relation to certain offences;
(k) restrict the punishment which may be imposed in respect of certain offences.

1.20 In addition to the above, Schedule 3 to HSWA contains detailed provisions about the contents of such Regulations, sufficient, it is thought, to completely replace the old law, and wide enough to enable the Secretary of State and the Commission to do almost anything in the interest of health and safety. In particular, however, we may note two further important provisions. The first is the power to prohibit the carrying on of any specified activity or the doing of any specified thing without a licence granted for that purpose, which may be subject to conditions. The second is contained in section 79 of the Act, which amends section 16 of the Companies Act 1967 so as to enable the Secretary of State to prescribe cases whereby directors' reports will contain such information about the arrangements in force for that year for securing the health, safety and welfare at work of the employees of that company (and any subsidiary company), and for protecting other persons against risks to health resulting from the activities at work of the employees. To date, no action has been taken to implement this provision.

1.21 A breach of a duty imposed by Regulations is, of course, punishable as a criminal offence. Additionally, a breach may give rise to civil liability, except in so far as the Regulations provide otherwise.

The making of Regulations

1.22 Health and Safety Regulations are made by the Secretary of State as a result of proposals made to him by the Health and Safety Commission (HSC). He may also make them on his own initiative, but before doing so he must consult with HSC and with any other appropriate bodies. If HSC makes the proposals, it too must consult with appropriate Government Departments and other interested bodies (HSWA, section 50). The normal practice is for HSC to initiate proposals in the form of a Consultative Document, which is given a wide circulation to interested bodies, such as employers' associations, trade unions, trade associations, and so on. Comments which are received are considered, and the proposals may be modified in the light of these. Draft Regulations are then publicised, amended if necessary, and the Regulations, in their final form, are presented to the Secretary for State, to be laid before Parliament.

1.23 Although Regulations are usually made by the Secretary of State for Employment, some have emanated from other Government Departments (e g Defence, Transport, Social Services).

C ORDERS

1.24 Historically, Orders were promulgated by the Queen in her Privy Council, exercising the Royal Prerogative. This procedure was a convenient way of avoiding Parliamentary scrutiny, but since the Statutory Instruments Act 1946 this is no longer so, and Orders are now subject to annulment on a resolution of either House of Parliament (see HSWA, section 84(4)). Modern practice is to confer on the appropriate Minister the power to make Orders, which again can only be exercised within the powers conferred. For example, section 84 of HSWA empowers the Queen by Order in Council to extend the provisions of the Act outside Great Britain (see Health and Safety at Work Act (Application outside Great Britain) Order 1977), whereas section 85 enables the Secretary of State to make Commencement Orders.

1.25 As a rough guide, the distinction between a Regulation and an Order is that the former is a means whereby the power to make substantive law is exercised, whereas the latter gives the force of law to an executive act.

D EUROPEAN LAW

1.26 By the European Communities Act 1972, section 2, Parliament accepted that the Treaty of Rome was to be a paramount source of law, which must be given priority over any contrary British rule (for example, see *Macarthys Ltd v Smith* for the effect of this ruling on the provisions of the Equal Pay Act 1970). In addition, the Council of Ministers has the power to issue Regulations and Directives, and the European Commission may give Decisions. A Regulation is directly enforceable in this country, and requires no further act of legislation by Parliament. A Directive, on the other hand, is normally an instruction to member states of the Community to pass legislation or to use other means as appropriate within national law to achieve proximity with the standards set out in the Directive. There is some support for the view that certain Directives are directly applicable without any need to pass legislation, but in view of the technical nature of health and safety standards, this is unlikely to be the case so far as Directives relating to occupational health and safety are concerned. A Decision is a particular ruling given in certain cases, e g where a state may request permission to depart from the terms of the Treaty. Questions of interpretation of Community law are reserved for the European Court sitting in Luxembourg, and its decisions must be followed by national courts (see chapter 11).

E BYELAWS

1.27 Local authorities have power to pass byelaws for the administration of their own area. These must be within the powers conferred by the parent Act of Parliament, and must also be reasonable. It is possible to challenge the validity of a byelaw in the courts.

F JUDICIAL PRECEDENT

1.28 The bulk of British law is contained in the decisions made by the judges in cases which come before them in the courts. When a judge decides a case, he will state his reasons for that decision. Frequently, the case will be reported in one of the many series of Law Reports (official or commercial) which exist. From the report, it may be possible to cull the narrow reason for the decision, i e the *ratio decidendi*. Anything else uttered in the course of the decision is regarded as *obiter dicta*, i e things said by the way. Distilling the *ratio* of a case, and distinguishing it from *obiter dicta*, is a legal art.

1.29 Strictly speaking, it is only the decisions of the higher courts which become binding precedents. Other decisions are said to be 'persuasive'. Thus a decision of the House of Lords will bind all lower courts, and can only be changed by a further decision of the House of Lords itself (a rare occurrence) or by an Act of Parliament. A decision of the Court of Appeal will also bind lower courts, but not, of course, the House of Lords. All other lower courts only create persuasive precedents.

1.30 However, lawyers will attempt to avoid inconvenient precedents (when necessary) by using their legal ingenuity. Thus it may be possible to distinguish a previous decision on the ground that the facts are not identical, or that there are material differences sufficient to warrant not following an earlier decision. Since the *ratio* of a case is uncertain, inconvenient remarks may be brushed aside as being *obiter dicta*, on the ground that they were not essential for the actual decision. A court may be asked to refuse to follow a precedent because it was decided 'per incuriam', i e without a full and proper argument on the point in issue. Frequently, there will be an abundance of precedents, and each side will quote those which support its argument. Occasionally, the point may never have arisen for decision before, and the court must reach its conclusions on first principles, thus creating a precedent. It is because of these points that the vagaries, as well as the richness, of British law emerges, but it will also be seen why litigation is so uncertain.

G APPROVED CODES OF PRACTICE

1.31 For the purpose of providing practical guidance, in recent years Parliament has authorised a number of organisations to issue Codes of Practice, which, though not having the force of law, may be taken into consideration by the courts and tribunals in appropriate circumstances. So far as this book is concerned, section 16 of HSWA conferred on the Health and Safety Commission power to approve and issue Codes of Practice for the purpose of providing practical guidance with respect to any of the general duties laid down in sections 2–7 of the Act, or any health and safety Regulations or existing statutory provision. The Commission may also approve other Codes which are drawn up by other persons or organisations (e g British Standards). The Commission cannot approve a Code without first obtaining the consent of the Secretary of State, and prior to obtaining this, it must consult with Government Departments and other appropriate bodies. Codes may be revised from time to time, and the Commission may, if necessary, withdraw its approval from a particular Code.

1.32 A failure on the part of any person to observe the provisions contained in an Approved Code of Practice does not, of itself, render that person liable to any criminal or civil proceedings, but in any criminal proceedings, if a person is alleged to have committed an offence concerning a matter in respect of which an approved Code is in force, the provisions of that Code are admissible in evidence, and a failure to observe it constitutes proof of the breach of duty, or contravention of the Regulation or statutory provision in question, unless the accused satisfies the court that he complied with the requirement of the law in some other equally efficacious manner. Approved Codes, therefore, are guides to good safety practice. If a person follows the requirements of the Codes it is unlikely that he will be successfully prosecuted for an offence. If he fails to follow the Code, he may be guilty of an offence unless he can show that he observed the specific legal requirements in some other way (HSWA, section 17).

1.33 The purpose behind the making of increasing use of Codes of Practice is to avoid built-in obsolescence in legal requirements, and to give practical guidance for the benefit of those upon whom duties are placed by virtue of a statute or Regulation. HSWA contains no guidance on the status of Codes of Practice in civil proceedings, but it is likely that a failure to observe their provisions could constitute prima facie evidence of negligence, which would have to be rebutted by evidence to the contrary.

H GUIDANCE NOTES

1.34 These are issued from time to time by the Health and Safety Commission or the Health and Safety Executive, and are mere expressions of opinion, with no legal force whatsoever. However, they are backed by considerable technical and scientific expertise, and should be regarded with great interest and given due weight, for they contain very useful practical advice. For example, the Guidance Notes issued along with the Safety Representative and Safety Committee Regulations (see chapter 2) are considerably more informative than the Code of Practice which accompanies those Regulations.

The judicial system in England and Wales

A MAGISTRATES' COURTS

1.35 These courts are presided over by Justices of the Peace, who are generally non-lawyers, appointed by the Lord Chancellor on the recommendation of a local Advisory Committee. The aim is to select magistrates from a wide cross-section of the local community, so that the Bench is a balanced one. Magistrates are advised on points of law by a (usually) legally qualified clerk, but it is the responsibility of the Bench to determine the facts and to make a decision. In certain large cities, full-time Stipendiary Magistrates may be found. These are qualified lawyers, and are paid a stipend. A stipendiary magistrate will usually sit alone, whereas lay magistrates will normally sit in benches of three.

1.36 A magistrates' court has some civil jurisdiction (mainly concerned with matrimonial disputes) but the bulk of the work is criminal, and indeed, over 94 percent of all criminal cases start and finish in the magistrates' courts. True, the overwhelming majority of these are motoring offences, but a fair amount of petty crime is also dealt with. The powers of punishment are normally limited to imposing a maximum fine of £1,000, or to send someone to prison for up to six months, although the relevant statute will usually contain the maximum penalties. Sometimes, in respect of continuing offences, a fine may be imposed for each day in which the offence is continued.

1.37 Procedurally, there are three types of offences. First, there is a summary offence, which can only be dealt with before a court of summary jurisdiction, i e the magistrates' court. Second, there is an indictable offence. This results in a formal document, known as an indictment, being drawn up. An accused will be brought before the magistrates for them to determine whether or not there is sufficient evidence (or a *prima facie* case) to commit the accused to the Crown Court for trial (hence these are known as committal proceedings) where the case will be heard before a judge and jury. Third, there are those cases which are triable either way (i e summarily or on indictment) depending on the option exercised by the accused and the gravity of the case, which may cause the magistrates to decide to send the matter for trial or to hear it themselves.

1.38 So far as health and safety legislation is concerned, some offences under HSWA, Factories Act, Offices, Shops and Railway Premises Act, etc are summary, others are triable either before the magistrates or at the Crown Court, depending on the seriousness of the offence.

1.39 The overlap between certain statutory provisions may cause some problems. For example, an obligation under the Factories Act may be an absolute one, e g to securely fence a dangerous part of any machinery, whereas under HSWA, an offence may be committed only if the accused failed to take steps which were reasonably practicable. In the former case, a failure to do the thing required is an offence, and the fact that it was difficult, or even impossible, to carry out the obligation is no defence, whereas in the latter case, the accused may be able to show that it was not reasonably practicable to do more than he in fact did.

1.40 If a person wishes to appeal from the decision of the magistrates' court on a point of law, this is done by way of a 'case stated' to the Queen's Bench Divisional Court. Either the prosecution or the defence may so appeal. Appeals against conviction and/or sentence may be made by the accused to the Crown Court, and will take the form of a re-hearing before a judge (without a jury).

B CROWN COURT

1.41 This court will hear indictable offences and those cases (usually of a more serious nature) which the prosecution or defence have elected to have tried before the Crown Court. Such cases are heard before a judge and jury. It is the function of the judge to ensure that the trial is conducted fairly, and to sum up for the jury on points of law. The jury returns a verdict depending on the facts. Under HSWA, some offences are punishable by an unlimited fine, and five particular offences may be punished by a term of imprisonment of up to two years. To date, there is no recorded case of anyone being imprisoned under the Act, and the highest recorded fine is in the region of £16,000. An appeal against conviction and/or sentence will lie to the Court of Appeal.

C DIVISIONAL COURT

1.42 The Queen's Bench Divisional Court is presided over by (usually) the Lord Chief Justice, who will normally sit with two other judges. This court will hear appeals by way of case stated from the decisions of the magistrates' courts.

D COUNTY COURT

1.43 This is the lowest civil court, having jurisdiction in cases where the claim for breach of contract or an action in tort does not exceed £5,000. Cases are heard before a County Court Judge. A recently introduced procedure enables 'small claims' to be taken before a Registrar.

E HIGH COURT OF JUDICATURE

1.44 For all practical purposes, this is the major civil court, and is presided over by a High Court Judge, sitting in the Queen's Bench Division (there are other specialised courts at this level which deal with commercial matters, admiralty claims, divorce, probate etc). Juries have long been abolished in all but the most exceptional civil cases, and thus the judge will be the sole arbiter of law and fact. This court is based in the Strand in London, but will also hear cases in various parts of the country. So far as this book is concerned, civil claims for industrial accidents etc are usually heard in this court.

F COURT OF APPEAL

1.45 This court acts as the Appeal Court from the decisions of the Queen's Bench in civil matters, the Queen's Bench Divisional Court in criminal matters, and from the Crown Court and County Court. It is presided over by the Master of the Rolls (at present Lord Denning) who sits with two other Lord Justices of Appeal. Appeals may only be made on a point of law, and thus cases will consist entirely of legal argument.

G HOUSE OF LORDS

1.46 In its legal capacity, the House of Lords is the highest court in the land. It consists of 'Law Lords', i e judges who have been 'promoted' from the lower courts. The Lord Chancellor (a political appointee) is entitled to sit, as is any ex-Lord Chancellor. Normally, the House will sit with five members present, but as, in theory, the House of Lords is part of the legislature, the judgments consist of 'speeches'. By convention, only the Law Lords and Lord Chancellors (past and present) will attend when the House is sitting in its legal capacity. Once a decision of the House of Lords is given on a point of law, it will bind all lower courts until the House gives a different ruling, or until the decision is changed by an Act of Parliament.

H INDUSTRIAL TRIBUNALS

1.47 These were created in 1964, and now have a wide jurisdiction relating to problems arising out of modern employment legislation. For the purpose of this book, industrial tribunals have three important functions, namely (a) to hear appeals from the decision of an inspector to impose a Prohibition or Improvement Notice; (b) to hear claims for unfair dismissal; and (c) to hear other claims arising out of the Employment Protection (Consolidation) Act 1978 (e g time off work for safety representatives) or other Acts of Parliament. An industrial tribunal consists of a legally qualified chairman, with two 'wingmen' chosen from each side of industry. Decisions of industrial tribunals, though not regarded as precedents, are valued for their reasoning and guidance. Any person may represent a party before an industrial tribunal, costs are not normally awarded to the winner (but see para 3.119), and proceedings are generally more informal than in other courts. An appeal from a decision of an industrial tribunal on a point of law relating to cases concerning Improvement and Prohibition Notices will go to the Queen's Bench Divisional Court, but appeals on all other matters will go to the Employment Appeal Tribunal.

I EMPLOYMENT APPEAL TRIBUNAL

1.48 This is a specialised court whose main function is to hear appeals on points of law from the decisions of industrial tribunals. It will consist of a High Court judge and two 'wingmen' chosen from each side of industry. Again, procedure is relatively more informal than in other courts, legal representation is not insisted upon, costs are not normally awarded to the winner, and the procedure is designed to hear cases with a maximum amount of simplicity and a minimum amount of delay. A further appeal may be made to the Court of Appeal.

The divisions of substantive law

1.49 All the laws of Great Britain may be found in the thousands of volumes of statutes and judicial decisions which line the shelves of law libraries. For convenience, we tend to subdivide subjects under different headings, each with their own special rules and application. Thus, if a person dies, interested parties will want to know something about the law of wills, or probate, or succession, as the case may be. If someone buys a house, he will get involved with land law, or conveyancing, and so on. In the field of occupational health, safety and welfare, we are mainly concerned with employment law, which is an amalgam of the law relating to contract, tort and crime. Each of these requires a brief examination.

A CONTRACT LAW

1.50 Most people enter into contracts almost every day of their lives. When a person buys a newspaper, goes on a train journey, eats a meal in a restaurant etc, he is entering into a contract. The law on this subject is made up largely of judicial decisions, supplemented by certain statutory provisions, and can be found in any recognised textbook on the subject. Occupational health and safety is fundamentally based on the existence of a contract of employment, and certain aspects of this will be dealt with in chapter 10. There is some force in the view that the duties which an employer owes to his employees to ensure their safety and health at work are based on contract (*Matthews v Kuwait Bechtel Corporation*, chapter 9).

B LAW OF TORT

1.51 A tort is a civil wrong, i e a wrongful act by one person which gives a right to the injured party to sue for a legal remedy, which is usually, but not exclusively, an action for damages. Torts are sub-classified under a number of well-known headings; thus defamation, trespass, nuisance, negligence, etc are all torts each with their own special rules and legal requirements. We will be concerned to a large extent with the tort of negligence, which is the breach of a legal duty to take care not to cause damage or injury to another. Thus an employer owes a duty to take care in order to ensure the health and safety of his employees (see chapter 9), as well as to others who may be adversely affected by the work activities.

C CRIME

1.52 A crime is an act which is punishable by the state in the criminal courts. Parliament has taken the view that employers who fail to comply with certain minimum standards shall be punished, in order to deter others from breaking the law. It cannot be pretended that the actual amount of the fines imposed acts as a great deterrent, for these seldom reflect the hazard created, nor the ability of the employer to pay. Nonetheless, the adverse publicity and the tarnishing of reputation is regarded as being undesirable. Modern management will take the view that health and safety is an objective to be pursued for its own sake, not through fear of punishment.

1.53 It is clear that the object of the law must be prevention, not retribution, rectification of potential hazards, not punishment for indifference. The criminal law should always be a weapon of last resort, rather than an instant reaction when offences are discovered. Certainly, this is the philosophy behind the HSE inspectorate, for the number of prosecutions undertaken will represent only a small fraction of the number of breaches which may be discovered.

The overlap of the law

1.54 Since there are no clear boundaries between legal concepts, an act may frequently have two or more legal implications. Thus an accident at work may result in a prosecution for a criminal offence; the same incident may result in a civil claim being brought for compensation. Indeed, in civil proceedings, the fact that there has been a criminal conviction in respect of the relevant incident is admissible in evidence (Civil Evidence Act 1968, section 11) for the purpose of proving that the defendant committed the offence, although it is not conclusive, and rebutting evidence may be given. The majority of legal decisions which explain and interpret the various statutory provisions arise out of civil proceedings, a fact which indicates how the law on occupational health and safety tends to work in practice.

Statutory interpretation

1.55 It is the task of the courts to interpret the words used by Parliament, not to legislate, although there can be little doubt in practice that by doing the former they achieve the latter. In recent years, the courts have taken a broader view in respect of reforming legislation, and sought to interpret the words by seeking the intentions of Parliament. An exploration of this aspect of judicial innovation is beyond the scope of this book.

1.56 All modern Acts of Parliament contain a section which deals with interpretation, and they will also define special terms where necessary. Thus, the meaning of the word 'factory' is contained in an extremely lengthy section in the Factories Act 1961 (see section

175). The Health and Safety at Work Act contains four sections (sections 52–53, 76, 82) which are concerned with the meaning of words and phrases used in the Act, and this practice is to be found in other relevant statutes and Regulations. It is to these that reference should be made in the first instance.

1.57 The Interpretation Act 1978 can be looked to for assistance in some circumstances. By that Act, unless the contrary is intended, a reference to a male includes a reference to a female, and vice versa, and a reference to the singular includes the plural. A 'person' includes a body of persons (whether corporate or unincorporate), 'month' means calendar month, and so on.

1.58 There are a number of rules for statutory interpretation which are well known and used in legal circles, sometimes with Latin tags. There is the 'literal rule', whereby the courts will look at the ordinary and grammatical sense of the words used. If this leads to absurdity or inconsistency, the 'golden rule' may be applied, whereby a construction is given which leads to a sensible interpretation, taking into account the whole statute. The tag 'ejusdem generis' (of the same species) may be used on occasions. For example, if a statute referred to 'any car, van, bus or other vehicle' it is doubtful if the latter phrase would include a bicycle, as the species is clearly motorised transport.

1.59 Sometimes, the courts may look to the interpretation already given to a particular word or phrase when used in another enactment, particularly when the two statutes are dealing with the same or similar subject matter. Thus words used in the Offices, Shops and Railway Premises Act may be construed on the same lines as identical words used (and judicially explained) in the Factories Act. This is known as construing 'in pari materia' (in similar circumstances).

1.60 The following are some of the words and phrases which are to be commonly found in the law on health and safety generally.

i *'Shall'; 'shall not'*

1.61 These words impose an absolute obligation to do (or not to do) the act or thing in question, and it is not permissible to argue that it is impracticable, difficult or even impossible to do it (or not to do). Thus, when the Factories Act, section 14 states that 'Every dangerous part of any machinery shall be securely fenced', it imposes an obligation to securely fence, and a failure to do so may result in criminal proceedings or, if someone is injured because of a failure to securely fence, a civil action for compensation. If it is impossible to use the machine when it is securely fenced, then the Act implicitly prohibits its use (*John Summers Ltd v Frost*) or the occupier of the factory uses it at his peril.

ii *'So far as is practicable'; 'best practicable means'*

1.62 This is a high standard, but not an absolute one. If something is practicable, it must be done; if the best practicable means are required, then it is up to the person on whom the obligation is placed to find the best practicable means, and to constantly bring his knowledge up to date. However, it is not practicable to take precautions against a danger which is not known to exist, although once the danger is known, it becomes practicable to do something about it (*Cartwright v GKN Sankey Ltd*). The test is one of foresight, not hindsight (*Edwards v National Coal Board*). The standard of practicability is that of current knowledge and invention (*Adsett v K and L Steelfounders and Engineers Ltd*). Once something is found to be practicable, it is feasible, and it must be done, no matter how inconvenient or expensive it may be to do it. The burden of proof to show that it was not practicable to do something, or to do more than was in fact done, is on the person upon whom the obligation is placed (in civil proceedings) and upon the defendant or accused in criminal proceedings (HSWA, section 40).

iii *'Reasonably practicable'*

1.63 This phrase causes more concern to practitioners than any other, although there is no magic or mystery about it. It should be looked at in the following way.

1.64 First, we start off with the presumption that if something is practicable, the courts will not lightly hold that it is not reasonably practicable (*Marshall v Gotham & Co*).

1.65 Second, it is a somewhat lesser standard than 'practicable'. It is permissible to take into account on the one hand the danger or hazard or injury which may occur, and balance it, on the other hand, with the cost, inconvenience, time and trouble which would have to be taken to counter it. As Asquith LJ stated in *Edwards v National Coal Board*, '"Reasonably practicable" is a narrower term than "physically possible" and seems to me to imply that a computation must be made by the owner in which the quantum or risk is placed in one scale and the sacrifice involved in the measures necessary for averting the risk (whether in money, time or trouble) is placed in the other, and that, if it be shown that there is a gross disproportion between them – the risk being insignificant in relation to the sacrifice – the defendants discharge the onus on them. Moreover, this computation falls to be made by the owner at a point in time anterior to the accident.' By way of example, we may cite *Marshall v Gotham & Co* where, although the roof of a mine had been tested in the usual way, it collapsed because of the existence of a rare geological fault. It was held that the employers were not liable for the injuries to the plaintiff which ensued. The court held that, 'The danger was a very rare one. The trouble and expense involved in the use of precautions, while not prohibitive, would have been considerable. The precautions would not have afforded anything like complete protection against the danger. . .'. In the circumstances, the employers had done all that was reasonable.

1.66 Third, in criminal proceedings, it shall be for the accused to prove that it was not reasonably practicable for him to do more than was in fact done to satisfy the duty or requirement (HSWA, section 40). However, he can establish this by the ordinary burden of proof, i e on the balance of probabilities.

1.67 Fourth, in the last analysis, it will be for the court to decide, as a question of fact, based on the evidence which can be adduced, whether or not something was reasonably practicable. In the last resort, therefore, the issue can only be determined after the event when it has been tested in court.

iv *'Relevant statutory provisions'*

1.68 This means the provisions of Part I of HSWA (including Health and Safety Regulations made under the Act), plus the following provisions, which are contained in Schedule 1 of HSWA:

	Provisions which are relevant statutory provisions
The Explosives Act 1875	The whole Act except sections 30 to 32, 80 and 116 to 121
The Boiler Explosions Act 1882	The whole Act
The Boiler Explosions Act 1890	The whole Act

	Provisions which are relevant statutory provisions
The Alkali, etc Works Regulation Act 1906	The whole Act
The Revenue Act 1909	Section 11
The Anthrax Prevention Act 1919	The whole Act
The Employment of Women, Young Persons and Children Act 1920	The whole Act
The Celluloid and Cinematograph Film Act 1922	The whole Act
The Explosives Act 1923	The whole Act
The Public Health (Smoke Abatement) Act 1926	The whole Act
The Petroleum (Consolidation) Act 1928	The whole Act
The Hours of Employment (Conventions) Act 1936	The whole Act
The Petroleum (Transfer of Licences) Act 1936	The whole Act
The Hydrogen Cyanide (Fumigation) Act 1937	The whole Act
The Ministry of Fuel and Power Act 1945	Section 1(1) so far as it relates to maintaining and improving safety, health and wellbeing of persons employed in and about mines and quarries in Great Britain
The Coal Industry Nationalisation Act 1946	Section 42(1) and (2)
The Radioactive Substances Act 1948	Section 5(1)(a)
The Alkali, etc Works Regulation (Scotland) Act 1951	The whole Act
The Fireworks Act 1951	Sections 4 and 7
The Agriculture (Poisonous Substances) Act 1952	The whole Act
The Emergency Laws (Miscellaneous Provisions) Act 1953	Section 3
The Mines and Quarries Act 1954	The whole Act except section 151
The Agricultural (Safety, Health and Welfare Provisions) Act 1956	The whole Act

	Provisions which are relevant statutory provisions
The Factories Act 1961	The whole Act except section 135
The Public Health Act 1961	Section 73
The Pipe-lines Act 1962	Sections 20 to 26, 33, 34 and 42, Schedule 5
The Offices, Shops and Railway Premises Act 1963	The whole Act
The Nuclear Installations Act 1965	Sections 1, 3 to 6, 22 and 24, Schedule 2
The Mines and Quarries (Tips) Act 1969	Sections 1 to 10
The Mines Management Act 1971	The whole Act
The Employment Medical Advisory Service Act 1972	The whole Act except sections 1 and 6 and Schedule 1

1.69 Since the passing of HSWA, many of the above provisions have been repealed or superseded in some way.

v *'Prescribed'*

1.70 If something is to be prescribed, it has to be specifically referred to at some future time. Thus, if a statute states that something shall be done 'in prescribed cases' then the power exists for someone (e g the Secretary of State) to prescribe the cases where that thing shall be done, and until the power is exercised, nothing has happened. For example, section 5 of HSWA imposes duties in relation to pollution control on persons having control of prescribed premises, but to date, no action has been taken to activate the section by prescribing the premises in question. If something is to be prescribed under HSWA, it will be done by Regulations (section 53(1)).

Burden of proof

1.71 In criminal cases, the general rule is that it is for the prosecution to prove its case – in the hallowed phrase – beyond reasonable doubt. However, in some circumstances, the burden of proof is on the defendant. For example, section 17 of HSWA requires the accused to show that he observed the specific statutory provision otherwise than by observing the Approved Code of Practice; section 40 of HSWA requires the accused to show that it was not practicable or not reasonably practicable to do more than in fact was done, etc. However, this burden can be discharged by proof on the balance of probabilities (*R v Carr-Briant*).

1.72 In civil cases, the rule is that the burden is upon the plaintiff to prove his case on the balance of probabilities. In this, he may be assisted by some rules of evidence.

i *Res ipsa loquitor*

1.73 If something is under the control of the defendant, and an accident occurs in circumstances such that it would not have happened unless there had been a want of care by the defendant, then a presumption is raised that the defendant has been negligent. The burden is then put on the defendant to explain the accident, and to show that there was no want of care on his part. In other words 'the facts speak for themselves'. However, the defendant may be able to show a convincing reason why he was not negligent (e g the accident was caused by the fault of a third party) in which case the burden of proof is thrown back to the plaintiff to prove his case in the usual manner.

ii *Shifting presumptions*

1.74 Sometimes, the mere statement of facts will give rise to an inferential presumption, which may cause the burden of proof to shift from one party to the other, requiring it to be rebutted by evidence. For example, in *Gardiner v Motherwell Machinery and Scrap Co Ltd*, the plaintiff sued in respect of dermatitis which he alleged was contracted at work. Lord Reid said, 'When a man who has not previously suffered from a disease contracts that disease after being subjected to conditions likely to cause it, and when he shows that it starts in a way typical of the disease caused by such conditions, he establishes a prima facie presumption that his disease was caused by those conditions.' Thus if a person contracts bronchitis after working in dusty conditions, or suffers deafness after being exposed to excessively noisy conditions, an evidential presumption arises that those conditions caused the disease or injury in question.

The scope of the law

1.75 For historic reasons, the United Kingdom is made up of four countries, namely, England, Wales, Scotland and Northern Ireland. Although there is a general common law operating throughout the United Kingdom, there are some procedural differences. A reference to Great Britain excludes Northern Ireland.

1.76 England and Wales may be regarded as one country for legal purposes, and no difference exists in the law or procedure (certain problems may arise over the use of the Welsh language, but this is irrelevant for our purposes). Scotland, on the other hand, retains its own legal system, which has certain distinctive terminology, procedure, and sometimes, different legal rules. As a general principle, the law on health and safety in Scotland is the same as in England and Wales, and will be so treated throughout this book. Acts of Parliament will usually state whether or not they apply to Scotland, and if they do not, it is frequently necessary to enact further legislation in order to bring the countries into harmony. Some Acts apply to England, Wales and Scotland, and merely note the different terminology by reference to the appropriate Scottish words. Northern Ireland, however, has its own legal system, though at the present time it is governed direct from the UK Parliament, and will usually follow the pattern already laid down for Great Britian. The Health and Safety at Work (Northern Ireland) Order 1978 created the Health and Safety Agency, with functions and responsibilities similar to those of the Health and Safety Commission.

1.77 By the Health and Safety at Work Act (Application outside Great Britain) Order 1977, the provisions of the Health and Safety at Work Act are extended to cover offshore oil and gas installations, pipeline work, offshore construction work, certain loading and unloading operations, and shipbuilding, repair work and diving operations which take place outside British territorial waters. This is additional to the provisions of the Mineral Workings (Offshore Installations) Act 1971, the Petroleum and Submarine Pipe-lines Act 1975 and Regulations made thereunder.

2 The institutions of health and safety

The role of Government

2.1 The prime responsibility for overseeing health and safety legislation is a political function of the relevant Government Department. The major role is played by the Department of Employment, the Head of which is a Secretary of State (with a seat in the Cabinet) assisted by various junior Ministers, one of whom will have been designated as having a special responsibility for occupational health and safety matters. Other problems of health and safety, which are partly concerned with the occupation, and partly of general concern to the whole community are dealt with by other Departments. Thus the control of emissions into the atmosphere comes under the aegis of the Department of the Environment, while the Department of Energy has overall responsibility for occupational safety on the UK's offshore oil and gas installations, as well as for the licensing of nuclear power stations.

2.2 The Secretary of State for Employment appoints the Chairman of the Health and Safety Commission, and not less than six nor more than nine other members. As to three of these appointees, he must first consult with such organisations representing employers as he considers appropriate, as to a further three, he must consult with trade union organisations, and as to the remainder, he must consult with organisations representing local authorities and other bodies concerned with matters of health, safety and welfare. He must also approve the appointment by the Health and Safety Commission of the Director of the Health and Safety Executive and of the two other members. He exercises a general control over the Commission by approving (with or without modifications) proposals submitted to him, and may give directions to the Commission with respect to its functions. He will make Regulations, and gives his consent to the issue of Codes of Practice by the Commission.

Health and Safety Commission (HSC)

2.3 This body was created by the Health and Safety at Work etc Act 1974 (HSWA), section 10, and has the prime responsibility for administering the law and practice on occupational health and safety. As already noted, it consists of a chairman and between six and nine members (currently there are only eight). Details of the terms of appointment and constitution of HSC can be found in Schedule 2 to HSWA.

General duties of HSC (section 1(1))

2.4 The general duty of HSC is to do such things and make such arrangements as it considers appropriate for the general purposes laid down in section 1(1) of HSWA, which are:

- (a) securing the health, safety and welfare of persons at work;
- (b) protecting persons other than persons at work against risks to health or safety arising out of or in connection with the activities of persons at work;
- (c) controlling the keeping and use of explosives or highly flammable or otherwise dangerous substances, and generally preventing the unlawful acquisition, possession and use of any such substance;
- (d) controlling the emission into the atmosphere of noxious or offensive substances from premises of any class prescribed.

Particular duties of HSC (section 11(2))

2.5 Additionally, HSC has the following duties:

- (a) assist and encourage persons concerned with matters relevant to any of the above general purposes to further them;
- (b) make such arrangements as it considers appropriate for the carrying out of research, the publication of results, the provision of training and information connected therewith, and to encourage research and the provision of training and information by others;
- (c) make arrangements for ensuring that government departments, employers, employees, employers' organisations, trade unions and others concerned are provided with an information and advisory service and are kept informed and adequately advised;
- (d) submit proposals for the making of Regulations by the appropriate authority (i e via the Minister responsible).

2.6 HSC shall also submit to the Secretary of State from time to time particulars of what it proposes to do for the purpose of performing its functions, it must ensure that its activities are in accordance with proposals approved by the Secretary of State, and give effect to any directions given by him.

Powers of HSC (section 13)

2.7 HSC may do anything (except borrow money) which is calculated to facilitate, or is conducive or incidental to, the performance of its functions, and in particular may

(a) enter into agreements with any Government Department or other person for that department or person to perform any function on behalf of HSC or the HSE. For example, HSC has made an 'agency agreement' with the Department of Transport whereby the latter operates on behalf of HSC in connection with the health, safety and welfare of persons who work in the operation, inspection or maintenance of statutory railway undertakings, as well as certain other activities, although the HSE retains the duty to inspect railway offices under the provisions of the Offices, Shops and Railway Premises Act (see chapter 5);

(b) make agreements whereby HSC performs functions on behalf of any other Minister, government department or other public authority, being functions which, in the opinion of the Secretary of State, can appropriately be performed by HSC (but not the power to make Regulations or other legislative instruments);

(c) provide services or facilities which may be required by any Government Department or public authority;

(d) appoint persons or committees to provide HSC with advice (see para 2.16);

(e) pay travelling and/or subsistence allowances and compensation for loss of remunerative time in connection with any of the functions of HSC;

(f) pay for any research connected with the functions of HSC and to disseminate (or pay for the dissemination of) information derived from such research;

(g) make arrangements for the making of payments to HSC by other parties who are using facilities and services provided by HSC.

Investigations and enquiries (section 14)

2.8 If there is an accident, occurrence, situation or other matter which HSC thinks necessary or expedient to investigate in order to fulfil any of the general purposes of the Act (above) or with a view to the making of Regulations for those purposes, it may direct HSE or any other person to carry out that investigation and make a report (section 14(2)(a)).

2.9 As an alternative, and with the consent of the Secretary of State, HSC may direct that an enquiry shall be held into the matter (section 14(2)(b)). The result of the investigation or enquiry may be made public, in whole or in part, as HSC thinks fit.

2.10 The distinction between an investigation and an enquiry is that an enquiry is a formal, public affair, into a matter in which there is general public interest as to its outcome, whereas an investigation is more informal, and may be carried out in a manner determined by the investigator, though he must still observe the rules of natural justice, and receive evidence from any person who may be able to contribute worthwhile information.

2.11 The Health and Safety Inquiries (Procedure) Regulations 1975 lay down the procedure for the conduct of enquiries under section 14(2)(b). The date, time and place shall be fixed (and may be varied) by HSC, who shall also give twenty-eight days notice to every person entitled to appear whose name and address is known. It must also publish a notice of the intention to hold the enquiry in one or more newspapers (where appropriate, circulating in the district in which the subject matter of the enquiry arose) giving the name of the person who has been appointed to hold the enquiry and of any assessors appointed to assist.

2.12 The following persons shall be entitled to appear at the enquiry as of right:

(a) the Commission;
(b) any enforcing authority concerned;
(c) in Scotland, the Procurator Fiscal;
(d) any employers' association or any trade union representing employers or employees concerned;
(e) any person who was injured or who has suffered damage as a result of the incident, or his personal representatives;

(f) the owner or occupier of any premises in which the incident occurred;
(g) any person carrying on the activity which gave rise to the incident.

Other persons may appear at the enquiry, but only at the discretion of the person appointed to hold it. Anyone wishing to appear at an enquiry may do so in person, or may be represented by a lawyer or by any other person.

2.13 The person appointed to conduct the enquiry may, either on his own volition or on the application of any person entitled or permitted to appear, require the attendance of any other person in order to give evidence or to produce any document, and it is an offence, punishable by a fine of up to £400 to fail to comply with such requirement.

2.14 The conduct of the enquiry will be the responsibility of the appointed person, and it will be held in public unless a Minister of the Crown directs that part or all of the enquiry shall be held in private because it would be against the interests of national security to allow certain evidence to be given in public. Also, a private session may be held if information is likely to be disclosed which relates to a trade secret, on an application made by a party affected. Otherwise, the procedure to be followed will be at the discretion of the appointed person, which he will state at the commencement of the proceedings, subject to any submissions made by any person appearing (or their representatives). He will also inform them as to any proposals he may have regarding an inspection of any site, will determine the order of the witnesses, permit opening statements, examination and cross-examination, hear evidence on oath, permit documents to be introduced as evidence and allow them to be inspected and copied. He may receive any written submissions, and if necessary, adjourn the proceedings from time to time.

2.15 After the enquiry has been concluded, a report, containing the findings of fact and any recommendations, will be made to HSC.

Advisory committees

2.16 A number of advisory committees have been created in order to provide HSC with specialised advice and information. These will usually consist of representatives from each side of industry, academic and industrial experts, and officials from HSE. They will report to HSC, and may make recommendations. Examples of existing Standing Advisory Committees are as follows:

i *Advisory Committee on Major Hazards*

2.17 This Committee was set up in order to investigate major hazards or actual disasters, and to make recommendations to minimise or eliminate the dangers from major incidents of the Flixborough type. It has invited views from various organisations on how to tackle the problems which arise from the existence of installations which represent potential major hazards, and issues reports for consideration.

ii *Advisory Committee on Toxic Substances*

2.18 This committee was set up to advise HSC on methods of controlling health hazards to persons at work and also to the general public consequent on or arising from the use of toxic substances. The committee has representatives from industry, local authorities and other expert advisers.

iii *Advisory Committee on Dangerous Substances*

2.19 This committee considers methods of securing the safety of persons at work and related risks to the general public arising from the manufacture, import, storage, conveyance and use of materials which are flammable or explosive, and the transportation of a wide range of dangerous substances. Its membership includes representatives from industry, local authorities and other experts.

iv *Medical Advisory Committee*

2.20 This committee considers the biomedical aspects of problems relating to occupational health, and includes in its work the identification of health hazards, biological monitoring, epidemiological studies, mental health and rehabilitation. As well as industrial representatives, the committee has a number of medical, nursing and academic members.

v *Nuclear Safety Advisory Committee*

2.21 This committee advises on all aspects of nuclear safety. It is chaired by an independent member, and has scientific and industrial advisers.

HSC/HSE outline main organisations chart at April 1980

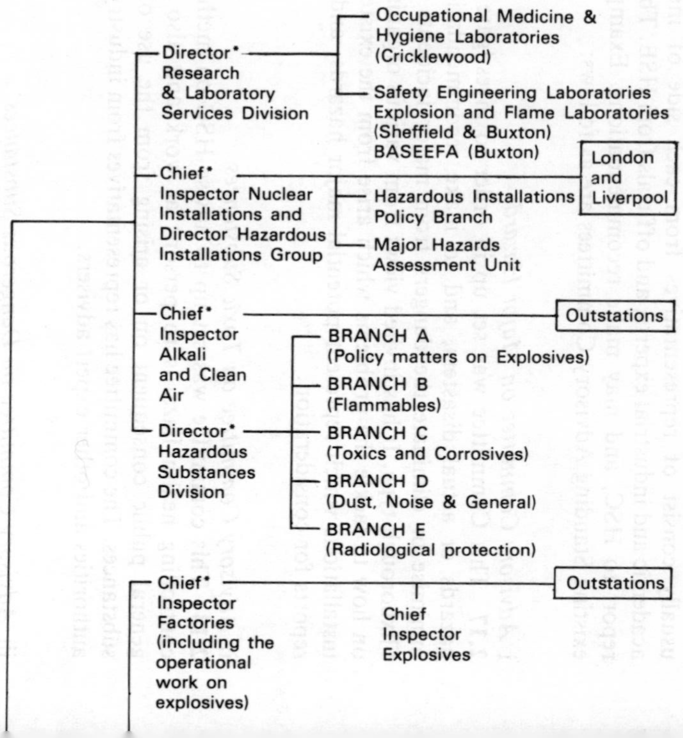
Director* Research & Laboratory Services Division
Occupational Medicine & Hygiene Laboratories (Cricklewood)
Safety Engineering Laboratories Explosion and Flame Laboratories (Sheffield & Buxton) BASEEFA (Buxton)
Chief* Inspector Nuclear Installations and Director Hazardous Installations Group
London and Liverpool
Hazardous Installations Policy Branch
Major Hazards Assessment Unit
Chief* Inspector Alkali and Clean Air
Outstations
BRANCH A (Policy matters on Explosives)
BRANCH B (Flammables)
Director* Hazardous Substances Division
BRANCH C (Toxics and Corrosives)
BRANCH D (Dust, Noise & General)
BRANCH E (Radiological protection)
Chief* Inspector Factories (including the operational work on explosives)
Outstations
Chief Inspector Explosives

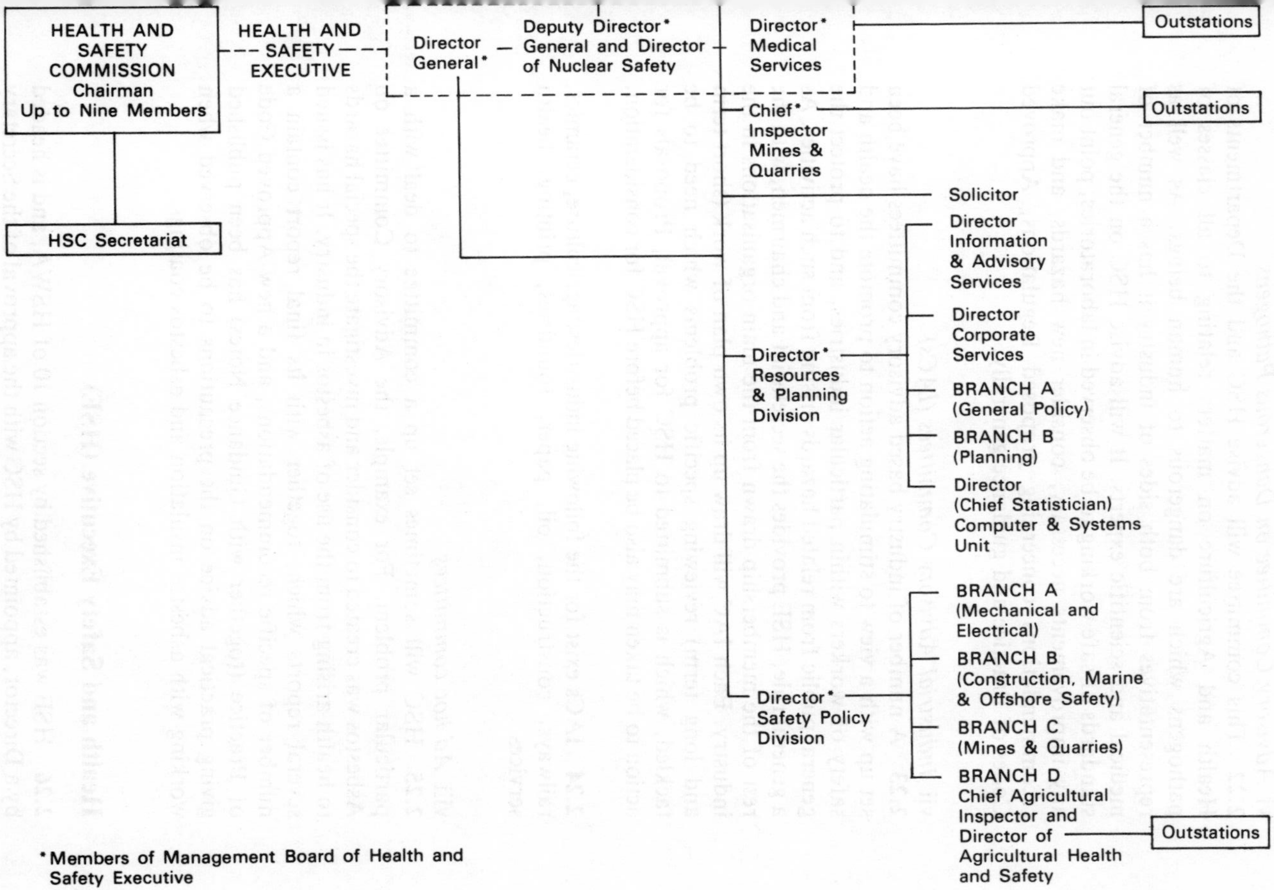
HEALTH AND SAFETY COMMISSION
Chairman
Up to Nine Members
HSC Secretariat
HEALTH AND SAFETY EXECUTIVE
Director General*
Deputy Director* General and Director of Nuclear Safety
Director* Medical Services
Outstations
Chief* Inspector Mines & Quarries
Outstations
Solicitor
Director* Resources & Planning Division
Director Information & Advisory Services
Director Corporate Services
BRANCH A (General Policy)
BRANCH B (Planning)
Director (Chief Statistician) Computer & Systems Unit
Director* Safety Policy Division
BRANCH A (Mechanical and Electrical)
BRANCH B (Construction, Marine & Offshore Safety)
BRANCH C (Mines & Quarries)
BRANCH D Chief Agricultural Inspector and Director of Agricultural Health and Safety
Outstations

*Members of Management Board of Health and Safety Executive

vi *Advisory Committee on Dangerous Pathogens*

2.22 This committee will advise HSC and the Departments of Health and Agriculture on matters relating to all classes of pathogens which are dangerous to human beings. As well as representatives from both sides of industry it has a number of medical and scientific experts. It will advise HSC on the general standards of safe working to be observed in laboratories, point out any improvements necessary, consider new hazards, and make recommendations concerning proposed Regulations, Approved Codes of Practice and guidance generally.

vii *Industrial Advisory Committees (IACs)*

2.23 A number of industry based advisory committees have been set up with a view to stimulating action to promote the health and safety of workers within particular industries, and to protect the general public from related hazards arising from such activities. As a general rule, HSE provides the secretariat and chairmen, with the rest of the membership drawn from the main organisations in the industry. Each IAC will draw up its own plan of work (short term and long term) reviewing specific problems which need to be tackled, which is submitted to HSC for approval, Proposals for action to be taken may also be placed before HSC for consideration.

2.24 IACs exist for the following industries: agriculture, ceramics, railways, construction, oil, paper, foundries, printing, health services.

viii *Ad hoc committees*

2.25 HSC will sometimes set up a committee to deal with a particular problem. For example, the Advisory Committee on Asbestos was created to consider and investigate the special hazards to health arising from the use of asbestos in industry. It has issued several reports which, together with its final report contain a number of specific recommendations, and a new Approved Code of Practice (together with Guidance Notes) has been published giving practical advice on the precautions to be observed when working with asbestos insulation and asbestos coatings.

Health and Safety Executive (HSE)

2.26 HSE was established by section 10 of HSWA, and is headed by a Director, appointed by HSC with the approval of the Secretary of State, and two other members, appointed by HSC with like approval, after consultation with the Director. HSE, which has its headquarters in London, controls the various branches of the Inspectorate, and operates through twenty-one area offices, each controlled by an Area Director.

2.27 The enforcement duties of the different branches are broadly speaking on the following lines:

(a) *Factory Inspectorate* has responsibility for all factories, construction work, and other premises formerly covered by the Factories Act, and all other employment covered by HSWA (including 'new entrants') except in so far as these areas are covered by other branches of the Inspectorate or come within the various agency agreements whereby responsibility is transferred to some other enforcement body (e g local authorities).

(b) *Explosives Inspectorate* has enforcement responsibilities for explosives factories, legislation dealing with compressed liquified gases (and other such substances) and the conveyance of petroleum, flammable gases and peroxides.

(c) *Mines and Quarries Inspectorate* is responsible for all health and safety matters in mines and quarries.

(d) *Nuclear Installations Inspectorate* is responsible for health and safety in nuclear power stations, radioactive fuel processing, research premises, the United Kingdom Atomic Energy Authority's premises, but premises which give rise to risks from ionising radiations (e g dentists' surgeries) are covered by the Factory Inspectorate.

(e) *Alkali and Clean Air Inspectorate* is principally concerned with the emission into the atmosphere of noxious or polluting substances from factories, quarries, etc, which come within the ambit of the Alkali etc Works Orders 1966 and 1971.

(f) *Agriculture Health and Safety Inspectorate* is responsible for health and safety in agriculture and kindred activities.

2.28 In addition, HSE controls the activities of the Employment Medical Advisory Service (see below). Thus, although industry has to deal with one unified body, in practice specialists from the different branches may be encountered, as well as other enforcement bodies (e g fire authority, local authorities, see below). Each inspectorate will obviously deal with those matters in respect of which it has been assigned.

National Industry Groups (NIGs)

2.29 The internal organisation of HSE is designed to permit a greater degree of industrial specialisation. This is done by means of National Industry Groups, which exist for the following industrial activities: breweries, ceramics, chemicals, construction, docks, electricity, food, footwear, foundries, engineering, higher education, hospitals, paper, plastics, printing, research establishments, rubber, shipbuilding, steel, textiles (wool and cotton), wire and rope-making, woodworking. In all, some twenty-one NIGs are in existence, each attached to an Area Office (see Appendix). Any inspector may approach the NIG specialist for assistance and information. In addition, a number of Area Offices have national responsibility for certain processes in respect of which NIGs do not exist.

Duties of HSE

2.30 It is the duty of HSE to exercise on behalf of HSC such of the latter's functions as it may care to direct that HSE shall exercise, and to give effect to any directions given to it by HSC. However, except for the purpose of giving effect to directions given to HSC by the Secretary of State, HSC cannot give HSE any directions as to the enforcement of the relevant statutory provisions in any particular case (section 11(4)). In other words, HSE alone will decide when and how to enforce the law in individual cases.

2.31 HSE may also provide a Minister of the Crown with information in connection with any matter with which he is concerned, and provide him with advice. HSE can do anything (except borrow money) which is calculated to facilitate, or is conducive or incidental to, the performance of any function within its statutory obligations. The duty of HSE to enforce the relevant statutory provisions is limited by the extent that some other authority is responsible for their enforcement (see Health and Safety (Enforcing Authority) Regulations 1977, below).

Enforcement by HSE

2.32 HSE has a number of actual and potential powers whereby it may seek to ensure compliance with the relevant statutory provisions. Primarily, however, it will seek to give advice and assistance to any employer or other person who is seeking ways of meeting the necessary standards. Persuasion, rather than compulsion, has always been the style adopted by the inspectorate, for the object has always been to establish high standards of health and safety, rather than to punish offenders. Thus an inspector may give verbal or written advice as to the steps which need to be taken to ensure compliance with the statutory duties. If remedial steps are not taken, or where the dangers are so obvious or imminent that immediate action is necessary, the new powers of issuing Improvement or Prohibition Notices may be exercised. As a last resort, criminal proceedings may be brought, which are usually instituted when there has been an incident, or after prior warnings have been given, or where the employer in question has a bad record of compliance. In other words, the inspectors will usually start off with advice, persuasion and encouragement. If these fail, resort may be had to compulsion or sanctions.

2.33 Every enforcement authority has the power to appoint as inspectors (under whatever title they may think fit) such persons having suitable qualifications as it thinks necessary for enforcing the relevant statutory provisions within its field of responsibility, and may also terminate the appointment (HSWA, section 19). However, it is submitted that this does not take inspectors outside the protection of the relevant provisions of the Employment Protection (Consolidation) Act 1978, particularly in respect of rights not to be unfairly dismissed. When an inspector is appointed, he will be given a document specifying which of the powers conferred on him by the relevant statutory provisions are to be exercisable by him, and he may only exercise those powers within the area of the responsibility of the enforcing authority which appointed him. When seeking to exercise those powers, he must produce, on request, his instrument of appointment. The powers of inspectors are as follows:

(a) at any reasonable time, or, if there is a dangerous situation, at any time, to enter premises which he has reason to believe it is necessary to enter for the purpose of carrying into effect any relevant statutory provision;

(b) to take with him a constable if he has reasonable cause to apprehend any serious obstruction in the execution of his duty;

(c) to take with him any person duly authorised by the inspector's authority, and any equipment or materials required for any purpose for which the power of entry is being exercised;

(d) to make such examination and investigation as may be necessary;
(e) to direct that any premises or anything therein shall be left undisturbed so long as it is reasonably necessary for the purpose of examination or investigation;
(f) to take such measurements and photographs and make such recordings as he considers necessary for the purpose of examination or investigation;
(g) to take samples of any article or substance found in any premises and in the atmosphere in or in the vicinity of any such premises; the Secretary of State may make regulations concerning the procedure to be adopted in such cases;
(h) in the case of any article or substance likely to cause danger to health or safety, to cause it to be dismantled or subjected to any process or test, but he may not destroy or damage it unless it is for the purpose of exercising his powers. If the person who has responsibilities in relation to those premises so requests, the inspector will exercise this power in that person's presence, unless he considers that it would be prejudicial to the safety of the state to do so. In any case, before exercising these powers, he must consult with appropriate persons for the purpose of ascertaining what dangers, if any, there may be in doing what he proposes to do;
(i) in the case of any article or substance likely to cause danger to health or safety, to take possession of it, and detain it for so long as is necessary in order to examine it, to ensure that it is not tampered with before he has completed his examination, or to ensure that it will be available for use as evidence in any proceedings for an offence, or in respect of matters arising out of the issuing of an Improvement or Prohibition Notice. He must leave a notice giving particulars of the article or substance, stating that he has taken possession of it, and, if it is practicable to do so, leave a sample with a responsible person;
(j) if he is conducting an examination or investigation under (d) above, to require any person whom he has reasonable cause to believe to be able to give any information to answer such questions as the inspector thinks fit, and to sign a declaration of the truth of his answers. However, no answer given in response to an inspector's questions shall be admissible in evidence against that person or his husband or wife in any proceedings;
(k) to require the production of, inspect, and take copies of, any entry in any books or documents which are required to be kept, and of any book or document which it is necessary

for him to see for the purpose of any examination or investigation under (d) above, but not if the production is refused on the ground of legal professional privilege;

(l) to require any person to afford him such facilities and assistance within that person's control or responsibilities, as are necessary for him to exercise his powers;

(m) any other power which is necessary for the purpose of exercising any of the above powers.

Obtaining information (section 27)

2.34 In order to discharge their functions, HSC, HSE or any other enforcing authority may need further information. To obtain this, power is given in section 27 to the HSC, with the consent of the Secretary of State, to serve on any person a notice requiring him to furnish HSC, HSE or any other enforcing authority with such information about such matters as may be specified in the notice, and to do this in such form and manner, and within such time, as may be specified.

2.35 No information obtained under this power shall be disclosed without the consent of the person by whom it was furnished, but this does not prevent the disclosure

(a) to HSC, or HSE, or Government Department or any other enforcing authority;

(b) by the recipient of the information to any person for the purpose of any function conferred on the recipient by any relevant statutory provision;

(c) to an officer of a local authority, water authority, or river purification board, or to a constable, in each case to a person who is authorised to receive it;

(d) by the recipient, as long as it is in a form which is calculated to prevent it from being identified as relating to a particular person or as a particular case;

(e) of any information for the purpose of any legal proceedings or any investigation or enquiry held by virtue of section 14(2) (above) or any report made as a result thereof.

Disclosure of information by inspectors (section 28)

2.36 A person exercising his powers of entry or inspection under section 14(4)(a), or his general powers of inspection etc under section 20 may not disclose any information obtained by him (including, in particular, any information with respect to any trade secret obtained by him in any premises into which he has entered by virtue of such powers) except

(a) for the purpose of his functions, or
(b) for the purpose of any legal proceedings, or any investigation or inquiry held by virtue of section 14(2), or a report thereof, or
(c) with the consent of the person who furnished the information in pursuance of a requirement imposed under section 20, and with the consent of the person having responsibilities in relation to the premises where the information was obtained in any other case (section 28(7)).

2.37 However, an inspector is expressly authorised by section 28(8) to disclose information to employed persons or their representatives if it is necessary for him to do so for the purpose of assisting in keeping them adequately informed about matters affecting their health, safety or welfare. Normally, this information will be communicated to safety representatives, but in the absence of these, it can be to a shop steward, or to individuals or even their legal representatives, should this be necessary. The information which can be disclosed is

(a) factual information obtained by him as a result of the exercise of any of his powers under section 20 (above) or as a result of a person who is holding an enquiry under section 14 (above) exercising his right to enter and inspect premises (see section 14(4)(a)), as long as the information relates to the premises where the employees are employed, or to anything which was or is in there, or was or is being done there, and
(b) information with respect to any action which he has taken or proposes to take in or in connection with those premises.

If the inspector does give such information to employees or their representatives, he must also give the like information to the employer.

Disclosure for civil proceedings (section 28(9))

2.38 Section 116 of the Employment Protection Act 1975 added a new subsection (section 28(9) to HSWA), which permits the disclosure by an inspector of information to a person who is likely to be a party to any civil proceedings arising out of an accident, occurrence or other situation. He will do this by providing a written statement of the relevant facts observed by him in the course of exercising any of his powers. Thus, in a civil action for damages (see chapter 9) either the plaintiff or the defendant may request the information from an inspector.

HSC has produced a Guidance Note on the subject.

Additional powers

2.39 As we shall see, an inspector may serve a Prohibition Notice, an Improvement Notice, and in England and Wales may prosecute offenders. He may seize and render harmless any article or substance likely to be a cause of imminent danger, and is responsible for the enforcement of the employer's duty to display the certificate of insurance under the provisions of the Employers' Liability (Compulsory Insurance) Act 1969.

Power of an enforcing authority to indemnify inspectors (section 26)

2.40 A person who is aggrieved by any action of an inspector who has exceeded his powers may always bring a civil action against him. If the circumstances are such that the inspector cannot legally claim the right of indemnity from the appointing authority, the latter may nonetheless indemnify him in whole or in part against any damages, costs, or expenses ordered to pay or incurred if the authority is satisfied that he honestly believed that he was acting within his powers and that his duty as an inspector required or entitled him to do the act in question.

2.41 An aggrieved person may also pursue a complaint against an inspector, or HSE or HSC through the machinery laid down in the Parliamentary Commissioner Act 1967 (which created the investigatory role of the Ombudsman).

Local authorities

2.42 One of the problems disclosed by the Robens Committee was the overlapping jurisdiction of the different enforcement authorities. Thus, a shoe repair shop – since it was both a factory and a shop – could be visited by the factory inspector and a local authority inspector, each concerned with his own responsibilities. To prevent this happening, section 18 of HSWA enables the Secretary of State, by Regulation, to make local authorities responsible for certain prescribed activities, to facilitate the transfer of responsibilities between HSE and local authorities, and to deal with uncertain areas.

2.43 For the purpose of enforcement of the relevant statutory provisions, a local authority is (in England and Wales) a district council, a London Borough Council, the Sub-Treasurer of the Inner Temple, the Under-Treasurer of the Middle Temple, or the Council of the Isles of Scilly. In Scotland, the enforcing local authority is an islands or district council. The enforcement functions are carried out by officials who have a number of different titles, although they are increasingly being known as Environmental Health Officers. (Details about how to obtain a Diploma in Environmental Health may be obtained from the Environmental Health Officers Association.) EHOs have all the powers of an HSE inspector (above).

2.44 The Health and Safety (Enforcing Authority) Regulations 1977 have drawn a clear line of responsibility between HSE and local authorities. Certain premises have been assigned to the latter, on the basis of the concept of 'the main activity' carried on there. These activities are:

(a) the sale or storage of goods for wholesale or retail distribution;
(b) office activities;
(c) catering services;
(d) the provision of residential accommodation;
(e) consumer services provided in shop premises (except dry cleaning and television repairs);
(f) dry cleaning in coin-operated units in laundrettes.

Exclusions

2.45 Local authorities do not have enforcement responsibilities for the following premises:

(a) retail or wholesale premises, being premises which are controlled by a railway undertaking, or are dock warehouses, container depots, water, sewerage, town or natural gas premises, premises used for the storage of minerals at mines or quarries, or storing petroleum spirit in garages, or used for the wholesale distribution of flammable substances;

(b) premises occupied or controlled by a local authority which is an enforcement authority;

(c) premises occupied or controlled by a local authority which is not an enforcement authority, e g county councils, Greater London Council;

(d) premises occupied or controlled by a police authority, the Receiver for the Metropolitan Police, or a fire authority;

(e) premises occupied by the UK Atomic Energy Authority;

(f) premises occupied by the Crown;

(g) premises in which motor vehicles are repaired or maintained.

Self inspection

2.46 On the principle that there shall be no self inspection, HSE will inspect the premises of the local authorities, and the local authorities will inspect those belonging to HSE (even though it is a Crown body). Although it might be expected that local authorities should be regarded as good employers and aware of their responsibilities, there have been several cases where they have been prosecuted by HSE and convicted of offences under HSWA. Clearly, this is an indication of human weaknesses within the organisation, rather than a wilful or neglectful disregard of the standards on health and safety.

Transfer of authority

2.47 Regulation 5 of the 1977 Regulations enables the responsibility for enforcement to be transferred in a particular case from HSE to a local authority, or from a local authority to HSE. This may only be done with the agreement of the enforcing authority which has current responsibility. Before such agreement is made, a notice of the proposal to make the transfer must be given to any person affected, who may lodge an objection within twenty-one days to HSC, and the transfer may only be made if HSC so directs. Once the transfer is made, the person affected shall be notified as soon as is reasonably practicable. Any proposal to transfer the responsibility for premises occupied or controlled by the Crown can only be done with the combined agreement of HSE, the local authority, and the Government Department concerned.

Cases of uncertainty

2.48 If there is any uncertainty in a particular case as to which authority is to be responsible for enforcement, HSE or a local authority may apply to HSC which, after considering the circumstances and taking into account any views expressed, including those of the occupier or controller of the affected premises, may assign the responsibility to whichever authority it thinks appropriate.

Liaison work

2.49 Representatives of HSE and local authorities have formed a liaison committee to discuss matters of mutual interest concerning the enforcement of standards of inspection, training, the recording and reporting of statistical information, etc. There are also liaison officers between the two bodies at a local level. Local authorities are obliged to perform the statutory functions imposed on them under HSWA, and to perform such functions as may be conferred, in accordance with such guidance as HSC may give to them from time to time (HSWA, section 18(4)).

Other matters

2.50 Under the Control of Pollution Act 1974 local authorities have powers relating to the control of noise from places of work, or building sites, etc, are empowered to enforce maximum noise standards, and to sanction remedial measures for the benefit of persons affected. Also, they are responsible for the control of noxious emissions into the air in respect of premises which are not registered under the Alkali Act.

Employment Medical Advisory Service (EMAS)

2.51 There has been a medical branch of the Factory Inspectorate since 1898, but the new streamlined medical service was created by the Employment Medical Advisory Service Act 1972, and it commenced operations in 1973. Its existence was continued by virtue of Part II of HSWA, and the Service is now an integral part of HSE.

2.52 The functions of EMAS are laid down in section 55 of HSWA, and are as follows:

(a) securing that the Secretary of State, HSC, the Manpower Services Commission, employers' organisations, trade unions, and occupational health practitioners can be kept informed and adequately advised on matters of which they ought to take notice, concerning the safeguarding and improvement of the health of persons who are employed or who are seeking emloyment or training for employment;
(b) giving to employed persons and persons seeking employment or training for employment information and advice on health in relation to such employment or training;
(c) other purposes of the functions of the Secretary of State relating to employment.

2.53 EMAS is headed by a Director of Medical Services, with Senior Employment Medical Advisers to be found throughout the various regions of the country. There is also a Chief Employment Nursing Adviser, and specialist advisers in toxicology, respiratory diseases, pathology research, and the medical aspects of rehabilitation. From its Head Office, EMAS controls a force of about 140 Employment Medical Advisors (EMAs) who are qualified registered medical practitioners and Employment Health Nurses (ENAs) who are similarly qualified.

2.54 In order to exercise its statutory functions, EMAS has responsibility for

(a) advising the inspectorate on the occupational health aspects of Regulations and Approved Codes of Practice;
(b) regular examinations of persons employed on known hazardous operations;
(c) other medical examinations, investigations and surveys;
(d) giving advice to HSE, employers, trade unions and others about the occupational health aspects of poisonous substances, immunological disorders, physical hazards (noise, vibrations, etc) dust, and mental stress, including the laying down of standards of exposure to processes or substances which may harm health;
(e) research into occupational health;
(f) advice on the provision of occupational medical, nursing and first aid services;
(g) advice on the medical aspects of rehabilitation and training for and placement in employment.

Powers of EMAS

2.55 An EMA or ENA has the same power of entry and investigation as inspectors have under section 20 (above) by virtue of an appointment as such. He can examine workers in those employments which Regulations require that they shall be examined at regular intervals (e g Control of Lead at Work Regulations 1980) but he cannot force a person to be examined against his will. Nor will he prescribe any treatment, but will refer the worker to his own family general practitioner.

2.56 Advice is also given on the medical aspects of employing young persons (i e below the age of eighteen). Employers are obliged to notify the local careers office whenever a young person is taken into employment, and this enables a check to be kept on those young persons who have a medical problem, and ensures that there is a continuous watch being kept on those who may need further medical supervision.

2.57 ENAs are an integral part of the Service, and will visit premises on request or as a result of a visit by an inspector, and give advice and assistance to employers, trade unions, safety officers, safety representatives, etc. They will also give advice to the Disablement Resettlement Officers and others about the advisability of employing someone who has a health problem.

2.58 Thus information about suspected health hazards will reach EMAS from a number of sources, enabling an investigation to take place, and advice given on how hazards can be reduced or eliminated. EMAS is frequently used as an advisory service, and is constantly engaged in research into health problems, in collaboration with HSE.

Safety representatives

2.59 One of the more important innovations to be found in HSWA is contained in section 2(4), which enabled the Secretary of State by Regulations to provide for the appointment by recognised independent trade unions of safety representatives from among the employees, who will represent them in consultation with the employers, and have other prescribed functions. It will then be the duty of every employer to consult with such representatives with a view to the making and maintenance of arrangements which will enable him and his employees to co-operate effectively in promoting and developing measures to ensure the health and safety at work of the employees, and in checking the effectiveness of such measures (section 2(6)).

2.60 The relevent Regulations have been made (Safety Representative and Safety Committee Regulations 1977) together with an Approved Code of Practice and non-statutory Guidance Notes. In addition, there is an Approved Code of Practice on Time Off for the Training of Safety Representatives.

APPOINTMENT OF SAFETY REPRESENTATIVES (SECTION 2(4))

2.61 A safety representative may be appointed by an independent trade union which is recognised by the employer for the purposes of collective bargaining. To be a trade union, it must be on the list of trade unions maintained by the Certification Officer under section 8 of the Trade Union and Labour Relations Act 1974 (TULRA). It will be an independent trade union if it has applied for and received a Certificate of Independence from the Certification Officer (issued if it can show that it is not under the domination or control of the employer, whether by way of the provision of financial benefits or otherwise) under section 8 of the Employment Protection Act 1975. It will be a recognised trade union if the employer recognises it for the purpose of negotiations relating to or connected with one or more of the matters specified in section 29(1) of TULRA 1974. There is no need to have a formal agreement concerning recognition; it is a question of fact, to be determined by the circumstances of each case as to whether or not the employer does recognise the trade union concerned (*NUTGW v Charles Ingram & Co Ltd*).

2.62 If an employer refuses to consult with the appointed safety representatives on the ground that he does not recognise the trade union concerned, it is likely that HSE will invoke the aid of the Advisory Conciliation and Arbitration Service (ACAS) to provide advice, but since the provisions of the Employment Protection Act 1975, sections 11–16 (which provided for a machinery whereby a trade union could obtain a formal declaration of recognition rights from ACAS) have been repealed (Employment Act 1980, section 19), there is no way this matter can be resolved without instituting proceedings in the magistrates' court for a breach of section 2(6).

2.63 The system whereby the legal right to appoint safety representatives is confined to trade unions has been the subject of some criticism. It is argued that it tends to perpetuate the divisions in industry between 'us' and 'them'. Safety, after all, should be the concern of all, be they employers, managers, employees, union officials, and so on, and it is wrong that safety representatives are to be seen against the backcloth of the battle ground of industrial relations. Against this, it is argued that that disputes which may arise over the functions and rights of a safety representative can more readily be resolved within the existing framework of collective bargaining machinery, and that to place the safety representative outside that machinery is to leave no avenue available for the resolution of disputes.

2.64 It should further be noted that in a non-unionised situation, there is nothing to stop an employer from creating his own system of safety representation, and although persons appointed will lack the statutory backing of HSWA and the Regulations, they can operate on a voluntary basis on the same lines as those who are union sponsored.

2.65 Although a trade union has the right to appoint safety representatives, there is no requirement that he should be a member of the trade union, or an active member. Again, this raises some argument in practice. Some trade unions will appoint shop stewards to be the safety representatives, on the ground that they are better trained, will not be overawed by management, and can handle the appropriate disputes machinery. On the other hand, this can cause some problems, for, as a shop steward, he may be subject to the processes of re-election, or there may be a conflict arising out of his functions as a shop steward and those he exercises as a safety representative. For example, if there are disciplinary proceedings being taken against an employee (see chapter 10) who has acted in breach of safety rules, the shop steward who has to represent his member will have to reconcile his actions with his belief that the safety rules must be upheld. To avoid such conflict, some trade unions will refuse to appoint shop stewards and look to other members to carry out this important work. This has the added advantage of spreading the workload and training opportunities, and involving more members in the activities of the union. Clearly, no single pattern will meet all the circumstances, and the matter must be regarded from a pragmatic viewpoint.

2.66 Neither the Regulations nor the Approved Code of Practice specify how many safety representatives should be appointed by the trade union concerned, and certain difficulties may well exist in multi-union situations. The Guidance Notes attached to the Regulations suggests that the appropriate criteria should be based on

(a) the total number of employees in the workplace;
(b) the variety of different occupations;
(c) the size of the workplace and the variety of workplace locations;
(d) the operation of shift systems;
(e) the type of work activity and the degree and character of inherent dangers.

2.67 However, the Regulations do state that so far as is reasonably practicable, a person appointed as a safety representative shall either have been employed by his employer throughout the preceding two years or shall have had at least two years experience in similar employment.

2.68 A person who has been appointed safety representative shall cease to be such when

(a) the trade union which appointed him notifies the employer in writing that the appointment has been terminated; or
(b) he ceases to be employed at the workplace. However, if he was appointed to represent employees at more than one workplace he shall not cease to be a safety representative so long as he continues to be employed at any one of them; or
(c) he resigns.

FUNCTIONS OF SAFETY REPRESENTATIVES

2.69 When an employer has been notified in writing by or on behalf of a trade union that a person has been appointed as a safety representative, and of the group of employees he is to represent, he shall have the right to be consulted by the employer with a view to the making and maintenance of arrangements which will enable the employer and his employees to co-operate effectively in promoting and developing measures to ensure the health and safety at work of the employees, and to check the effectiveness of those measures (HSWA, section 2(6)). In addition, regulation 4 lays down a number of other functions which a safety representative shall be entitled to perform, as follows:

(a) to investigate potential hazards and dangerous occurrences at the workplace (whether or not they are drawn to his attention by the employees he represents) and to examine the causes of accidents at the workplace. This right is not confined to having time off work (with pay) to make the investigation inside the workplace, for there may be circumstances when it is necessary to go outside in order to investigate (e g to interview an injured employee at home, see *Dowsett v Ford Motor Co*, para 2.85);

(b) to investigate complaints by any employee he represents relating to that employee's health, safety or welfare at work;
(c) to make representations to the employer on the above matters;
(d) to make representations to the employer on general matters affecting the health, safety or welfare at work of the employees at the workplace;
(e) to carry out inspections (see below);
(f) to represent employees in consultations at the workplace with inspectors of HSE or any other enforcing authority;
(g) to receive information from inspectors (see para 2.37);
(h) to attend meetings of the safety committee where he attends in his capacity as safety representative in connection with any of the above functions.

2.70 As an employee, the safety representative is subject to the general requirements imposed by sections 7–8 of HSWA (see chapter 3) but no function conferred on him as noted above shall impose any duty on him. This means that he cannot be prosecuted for a failure to perform his duty as a safety representative, or for performing his duties badly. Nor can he be sued civilly for a failure to perform his statutory duty, although he is subject to the ordinary law of negligence in the usual way (see chapter 9).

INSPECTION OF THE WORKPLACE

2.71 Safety representatives are entitled to inspect the workplace or any part of it on three occasions.

2.72 First, if they have not inspected it within the previous three months. They must give reasonable notice to the employer in writing of their intention to do so. More frequent inspections may be carried out with the agreement of the employer.

2.73 Second, where there has been a substantial change in the conditions of work (whether by way of the introduction of new machinery or otherwise), or where new information has been published by HSC or HSE relevant to the hazard since the last inspection, then, after consulting with the employer, a further inspection may be carried out notwithstanding that less than three months have elapsed since the last inspection.

2.74 Third, where there has been a notifiable accident or dangerous occurrence (see chapter 6) or a notifiable illness has been contracted there, and

(a) it is safe for an inspection to be carried out, and
(b) the interests of the group of employees represented by the safety representatives might be involved.

2.75 In these circumstances, the safety representatives may carry out an inspection of the part of the workplace concerned (and so far as is necessary to determine the cause, they may inspect any other parts of the workplace). Where it is reasonably practicable for them to do so, they shall notify the employer of their intention to carry out the inspection.

2.76 The employer shall provide such facilities and assistance as the safety representatives may reasonably require, including facilities for independent investigation by them and private discussions with employees, but the employer or his representative is entitled to be present in the workplace during the inspection.

INSPECTION OF DOCUMENTS

2.77 For the performance of their duties, safety representatives are entitled, on giving reasonable notice to the employer, to inspect and take copies of any document relevant to the workplace or to employees they represent which the employer is required to keep by virtue of any relevant statutory provision. Thus, they can inspect and take copies out of the General Register, reports of the examination of hoists, lifts, cranes, steam boilers, etc. However, they are not entitled to see a copy of the Fire Certificate issued under the Fire Precautions Act (see chapter 6) as this is not one of the relevant statutory provisions. Nor are they entitled to inspect or take copies of any document consisting of or relating to any health record of an identifiable individual.

DISCLOSURE OF INFORMATION TO SAFETY REPRESENTATIVES

2.78 Safety representatives will be entitled to receive information from two main sources. First, regulation 7(2) requires the employer to make available information within his knowledge which is necessary to enable the safety representatives to perform their functions. However, the employer need not disclose

(a) information the disclosure of which would be against the interests of national security;
(b) any information which he could not disclose without contravening a prohibition imposed by or under an enactment;
(c) any information relating specifically to an individual, unless he has consented to it being disclosed;
(d) any information, the disclosure of which would, for reasons other than its effect on health, safety or welfare at work, cause substantial injury to the employer's undertaking, or, where the information was supplied to the employer by some other person, to the undertaking of that other person;
(e) any information obtained by the employer for the purpose of bringing, prosecuting or defending any legal proceedings.

2.79 Nor do the Regulations require the employer to disclose any document or to allow the inspection of it, if it does not relate to health, safety or welfare at work.

2.80 The restriction concerning the discovery of documents used for the purpose of legal proceedings was an issue before the House of Lords in *Waugh v British Railways Board* (see chapter 9) where it was held that for a document to be privileged, the dominant purpose in preparing it must have been for its use in possible litigation. Thus, if an accident report is prepared as a matter of routine practice in order to establish the cause, and is subsequently required for the purpose of litigation, the safety representative will be entitled to see it, for the dominant purpose in preparing it was not for the purpose of litigation.

2.81 The Approved Code of Practice makes a number of recommendations concerning the nature of the information which should be disclosed to safety representatives. These include:

(a) information about the plans and performance of the undertaking and any changes proposed in so far as they affect the health and safety at work of employees;
(b) information of a technical nature about the hazards to health and the precautions deemed necessary to eliminate or minimise them, in respect of plant, machinery, equipment, processes, systems of work, and substances in use at work,

including any relevant information provided by consultants and designers, or the manufacturer, importer or supplier of any article or substance used at work;

(c) information which the employer keeps relating to the occurrence of any accident, dangerous occurrence or notifiable disease, and any statistical records relating to those matters;

(d) any other information specifically relating to matters affecting health and safety at work, including the results of any measurements taken by the employer (or person acting on his behalf) in the course of checking the effectiveness of his health and safety arrangements;

(e) information on articles or substances which an employer issue to his homeworkers.

2.82 The second source of information for safety representatives may come from the inspector, who, by virtue of section 28(8) of HSWA may give information to employees or their representatives to ensure that they are adequately informed about matters affecting their health, safety or welfare (see above, para 2.37). This may be factual information relating to the premises or anything therein, or information regarding any action taken or proposed to be taken by the inspector (e g the issuing of a Prohibition or Improvement Notice). The inspector must give the like information to the employer.

TIME OFF WORK FOR SAFETY REPRESENTATIVES

2.83 A safety representative is entitled to have time off work, with pay, during his working hours, for the purpose

(a) of performing his functions as a safety representative;

(b) to undergo training in aspects of those functions as may be reasonable having regard to the Approved Code of Practice issued by HSC.

2.84 If his pay does not vary with the amount of work done, then he shall be paid as if he had worked throughout the whole of the time. If his pay varies with the amount of work done, then he shall be entitled to be paid his average hourly earnings for his work, or, if no fair estimate of his earnings can be made, the average hourly earnings for work of that description of persons in comparable employment, or, if there are no such persons, then the average hourly earnings which are reasonable in the circumstances.

2.85 The right to investigate potential hazards and dangerous occurrences is not confined to having time off (with pay) to make an investigation inside the workplace. In *Dowsett v Ford Motor Co* an employee was injured in an accident. The applicant, who was the safety representative, investigated the incident, and concluded that no further action was necessary. Five weeks later, he attended a meeting of the Works Safety Committee where the safety engineer gave a report on the incident. The applicant wanted to have time off work with pay to visit the employee at home in order to make further enquiries, but this was refused, and he made an application to the industrial tribunal. It was held that the Regulations were sufficiently wide to enable the safety representative to go outside the workplace if it was necessary to perform his functions. However, this was a question of fact and degree. In this case, he had not done anything for five weeks, and would not have acted had he not heard the report at the Safety Committee. Consequently, it was not necessary to make any further enquiries. In principle, however, the industrial tribunal made it quite clear that there could be circumstances where it would be necessary to interview the injured person (or others) and to perform this function satisfactorily it may be necessary to go outside the workplace.

TRAINING OF SAFETY REPRESENTATIVES

2.86 The Approved Code of Practice on 'Time off Work for the Training of Safety Representatives' states that as soon as possible after they have been appointed they should be permitted to have time off work to attend basic training facilities approved by the TUC or their own trade union. Further training, similarly approved, should be undertaken when they have special responsibilities, or when this is necessary because of changed circumstances or new legislation. The trade union should inform management of the course it has approved and supply a copy of the syllabus if the employer asks for it. The trade union should give a few weeks notice, and the number of safety representatives attending at one time from the same employer should be that which is reasonable in the circumstances, bearing in mind the availability of the relevant courses and the operational requirements of the employer. Unions and management should endeavour to reach agreement on the appropriate numbers and arrangements, and refer any problems to agreed procedures.

2.87 It will be recalled that the status of the Code of Practice is not of a rule of law; it is guidance to good practice. Thus there is no absolute rule that training should be only on a union approved course. In *White v Pressed Steel Fisher Ltd* the applicant was appointed safety representative by the T & GWU. The union wanted him to go on a union-sponsored training course, but management wanted to provide an in-company course, and refused to permit him to have time off work with pay to attend the union course. It was held by the EAT that the employers were not acting unreasonably in refusing him time off work to go on the union course. The provisions in regulation 4(2) were for such training as may be reasonable in all the circumstances. It was therefore necessary to consider all the circumstances, including the Code of Practice. The approval of a course by the trade union was a factor to be taken into account, but (unlike the provisions in the Employment Protection (Consolidation) Act 1978, section 27 - which relates to time off work for training in trade union duties) the Safety Representative and Safety Committee Regulations do not require that the course must be approved by the TUC or by the trade union. If the course provided by the employer was adequate, and contained all the necessary material, including the trade union aspects of safety, then it could be perfectly proper for the employer to insist that the safety representative went on the in-house course.

2.88 The Code of Practice also gives guidance on the contents of a safety training syllabus. Basic training should provide for an understanding of the role of a safety representative, of safety committees, and of trade union policies and practices in relation to

(a) the legal requirements relating to health and safety at work;
(b) the nature and extent of workplace hazards, and of the measures necessary to eliminate or minimise them;
(c) the health and safety policy of the employer, and the organisations and arrangements necessary to fulfil these policies.

2.89 In addition, they will need to develop new skills, including how to carry out a safety inspection, how to make use of basic sources of legal and official information, etc.

NON-EMPLOYEES AS SAFETY REPRESENTATIVES

2.90 The Regulations specify that if the safety representatives have been appointed by the British Actors' Equity Association or the Musicians Union, it is not necessary that they shall be employees of the employers concerned. This is because such people are performers in theatres etc which are not owned by their employer, and are generally itinerant workers.

Safety committees

2.91 The Safety Representatives and Safety Committee Regulations lay down (regulation 9) that the employer must establish a Safety Committee if requested to do so in writing by two safety representatives. In order to do this, he must consult with safety representatives who made the request and with the representatives of any recognised trade union whose members work in the workplace in respect of which he proposes to establish the committee. The duty is one of consultation, not negotiation or agreement, so that actual composition of the committee is a matter for the employer to determine. However, he must establish the committee within three months of the request. He must also post a notice stating the composition of the committee and the workplace covered by it, and the notice shall be posted in a place where it may easily be read by the employees.

2.92 Section 2(7) of HSWA states that the function of the safety committee shall be to keep under review the measures taken to ensure the health and safety at work of employees and such other functions as may be prescribed. Apart from this vague generalisation, neither the Regulations nor the Code of Practice give any further indication as to its functions. However, the Guidance Notes give some very helpful information. HSC believe that the detailed arrangements necessary will evolve from discussions and negotiations between the parties, who are best able to determine the needs of the particular workplace. Since the circumstances of each case will vary a great deal, no single pattern is possible.

2.93 Certain guides may be followed. The safety committee should have its own separate identity, and not have any other function or tasks assigned to it. It should relate to a single establishment although group committees can play an additional role. Finally, its functions should be clearly defined. A suggested brief for the committee might go along the following lines:

(a) a study of the trends of accidents, dangerous occurrences and notifiable diseases, so that recommendations may be made to management for corrective action to be taken;
(b) the examination of safety audit reports, to note areas where improvements can be made;
(c) consideration of reports and factual information from the enforcing authority;

(d) the consideration of reports made by the safety representatives;
(e) assisting in the development of safety rules and safe systems of work;
(f) an evaluation of the effectiveness of the safety content of employee training;
(g) monitoring of the adequacy of health and safety communication and publicity;
(h) acting as a link between the company and the enforcing authority;
(i) evaluating the safety policy and making recommendations for its revision.

Safety committees are not specifically empowered to deal with welfare matters, though there is nothing to prevent this happening should the parties so decide.

MEMBERSHIP OF THE COMMITTEE

2.94 The aim should be to keep the membership reasonably compact, with adequate representation from all interested parties. Management representatives should include persons involved in health and safety matters, e g works engineers, works doctor, safety officer, etc, and there should be seen to exist some form of mechanism for the consideration and implementation of the recommendations by senior management. The committee must contain sufficient expertise to evaluate problems and come up with solutions. Outside specialists may be made ex-officio members. There is no requirement that safety representatives should be members of the committee, but it would clearly be desirable to håve some representation, depending on the numbers involved. There is no provision for time off work to be paid for attendance at the meetings of the committee, but this should be obvious. It is also desirable that a senior member of the company attends the meetings and plays a leading role, e g the company chairman or a board director.

MEETINGS

2.95 The safety committee should meet on a regular basis, as frequently as necessary, depending on the volume of business. The date of the meetings should be promulgated well in advance, and provision should be made for urgent meetings. An agenda should be drawn up, minutes kept, with action taken noted. Probably the most important function of the committee will be to monitor action taken on any recommendations made.

SAFETY COMMITTEES IN NON-UNIONISED WORKPLACES

2.96 HSC has issued some guidance on safety committees in those premises where there is no recognised independent trade union. Since there is no formal trade union machinery, and no legal requirement to set up a committee, the initiative will presumably come from management. Again, there should be adequate representation of appropriate management skills, and the employee representatives should be chosen by their fellow employees. In cases of difficulty, HSE will give further guidance.

Enforcement of the Regulations

2.97 A safety representative may present a complaint to an industrial tribunal that

(a) the employer has failed to permit him to take time off for the purpose of performing his functions as a safety representative, or to permit him to go on a training course, or
(b) the employer has failed to pay him for his time off.

2.98 The complaint must be presented within three months from the date when the failure occurred or, if it was not reasonably practicable to do so within that time, within such further period of time as the industrial tribunal considers to be reasonable. If the industrial tribunal upholds the complaint that the employer has failed to permit him to have time off, a declaration to that effect shall be made. Additionally, the industrial tribunal may make an award of compensation to be paid by the employer to the employee, which shall be of such amount as the tribunal considers to be just and equitable in all the circumstances, having regard to the employer's default in failing to permit the employee to have the time off, and to any loss suffered by the employee as a result of that failure. The compensation awarded would normally be a modest amount, for it will be rare that an employee actually suffers any loss. In *Owens v Bradford Health Authority* a trade union appointed as a safety representative a man who was within a year of retirement. He applied to go on a training course, but the employers refused, as they did not think it was reasonable for a safety representative so near to retirement to go on a training course. The industrial tribunal upheld the employee's complaint, and awarded him £50 compensation. It is the prerogative of the trade union to make the appointment, and a refusal to send someone on a training course because he was near to retirement was not justified.

2.99 If the complaint is that of a failure to pay for the time off work, and it is upheld by the industrial tribunal, an award of the amount due to be paid will be made.

2.100 The enforcement of the other duties in the Act and the Regulations is the responsibility of the appropriate enforcing authority. Thus if the employer fails to consult with the safety representatives, as required by section 2(4), or fails to set up a safety committee as required by section 2(7), or fails to provide the necessary information as required by regulation 7, he will be in breach of the law and thus commits a criminal offence. HSC has issued guidance notes on how HSE should attempt to deal with the enforcement of these matters. In three cases, they will not go to immediate enforcement, but will try other means. These are:

(a) HSE must be satisfied that all voluntary means have been explored, including the taking of advice from or using the services provided by the Advisory, Conciliation and Arbitration Service (ACAS);

(b) if there is some doubt as to whether or not the trade union has made a valid appointment as a safety representative. For example, the employer may allege that the trade union is not recognised by him. This problem might be resolved by using the services of ACAS;

(c) if the problem relates to time off work, or a failure to pay, the specified remedy of using the machinery of the industrial tribunal should first be explored.

2.101 Enforcement by HSE might be appropriate in other cases, for example:

(a) if the trade union or safety representative complain that the employer is not carrying out his obligations after full use has been made of any consultative machinery or disputes procedure;

(b) where the employer is refusing to acknowledge the existence of the safety representative who has been validly appointed;

(c) where the employer refuses to make particular information available, or refuses to provide particular facilities to enable the safety representative to perform his functions;

(d) where the employer has refused or failed to set up a safety committee after being requested to do so under regulation 9.

2.102 Since the employer will be in breach, and is 'contravening one or more of the relevant statutory provisions' it is possible that instead of prosecuting for such breaches, the inspector may issue an Improvement Notice under section 21. To date, this tactic does not appear to have been necessary. Indeed, it is a credit to all concerned that the implementation of the provisions relating to safety representatives has, despite original fears, proceeded with great smoothness.

Safety officers

2.103 There is no general legal requirement to appoint safety officers (under whatever title) although as a matter of good practice many industrial and commercial organisations do so. There is a requirement to appoint safety supervisors or competent persons to ensure compliance with statutory requirements under certain Regulations, e g Clay Works Regulations 1948, Pottery (Health and Safety) Special Regulations 1950, Construction (General Provisions) Regulations 1961, Ship Building and Ship-repairing Regulations 1960, Ionising Radiations (Unsealed Radioactive Substances) Regulations 1968, and Ionising Radiations (Sealed Sources) Regulations 1969. No formal standards of training are laid down, the obligation is merely to ensure that he has sufficient time to discharge his duties efficiently, have experience and expertise necessary to carry out those duties, and have the authority to perform his duties.

2.104 There are also a number of statutory provisions which require an employee to be 'qualified' or 'trained', and other provisions require that certain things may only be done by or under the supervision of a 'competent person'. These phrases are always left undefined, and it may well be that the burden is on the employer to show in any given case that the person who performed the task in question was qualified, trained or competent, as the case may be. The possession of some formal certificate or qualification would no doubt assist in discharging this burden, but it may also be shown that the person concerned has pursued some approved course of training or instruction, as well as possessing practical experience of the work.

2.105 The existence of a safety officer, with details of his functions and powers, would be one of the things an employer would refer to in the written statement of his general policy on health and safety at work, as required by section 2(3) of HSWA, being part of the organisation and arrangements in force for the carrying into effect of that policy.

2.106 The relevant professional organisation for safety officers is the Institute of Occupational Safety and Health (IOSH), which was formed in 1953, and has about 3,000 members. It runs its own examinations, leading to Associate and Membership grades. The syllabus includes (a) law, (b) techniques of safety management, (c) behavioural science, (d) occupational health and hygiene, and (e) general science. There is also the Institute of Municipal Safety Officers, which has about 1,000 members, and a merger between the two organisations is currently under discussion.

2.107 HSE has produced a useful Discussion Document on the role and functions of safety officers which is available on request.

Occupational health services

2.108 Although there is no general legal obligation to provide medical services at the place of work (but see Health and Safety (First-Aid) Regulations 1981, chapter 6) many employers engage trained medical personnel to ensure the immediate treatment of injuries or illnesses which may occur during the employment, carry out pre-employment screening, investigate existing or potential medical hazards, or generally to provide a health care service for the benefit of employees. Occupational health services, by their very nature, are usually found in large firms, although there is a recent trend for smaller firms to pool together their resources and run a joint scheme. Sometimes the impetus comes from the need to reduce the incidence of accidents or ill-health, and consequently reduce the number of days lost through absenteeism. In other cases, the service may be part of the company's philosophy to provide additional welfare services on the premises for the use of all employees. Whatever the motive, occupational health services are on the increase.

2.109 Doctors who practice in industry may obtain a Diploma in Occupational Health or an M Sc degree in occupational medicine. State Registered Nurses may obtain the Occupational Health Nursing Certificate.

2.110 The relationship between works doctors (or nurses), management, trade unions, the individual and his own family doctor is a complex one, for problems of confidentiality and conflicting interests may arise. The Royal College of Physicians has issued a booklet *Guidance on Ethics for Occupational Physicians*, and the Royal College of Nursing has issued a list of duties which may be expected to be performed by an occupational nurse.

Other institutions

ROYAL SOCIETY FOR THE PREVENTION OF ACCIDENTS

2.111 RoSPA is the largest independent safety organisation in Europe. A major part of its activities is concerned with occupational health and safety, and it publishes three journals on the subject: (1) *Occupational Safety and Health*, (2) *RoSPA Bulletin*, (3) *Safety Representative*. RoSPA also organises conferences and training courses, and gives advice and assistance to members. Its address is Cannon House, The Priory, Queensway, Birmingham B4 6BS.

BRITISH SAFETY COUNCIL

2.112 A non-profit-making independent body, financed entirely by subscriptions from its members, the BSC provides information and training on all aspects of health and safety, will undertake loss control surveys, issues posters and booklets, runs a National Safety Award scheme, and issues a Diploma in Safety Management. BSC also publishes a magazine *Safety*. The address is 62/64 Chancellor's Road, Hammersmith, London W6 9RS.

BRITISH STANDARDS INSTITUTE

2.113 The BSI was incorporated by Royal Charter in 1929, and has as its objectives:

(a) to co-ordinate the efforts of producers and users for the improvement, standardisation and simplification of engineering and industrial materials so as to simplify production and distribution, and eliminate national waste of time and material in the production of unnecessary varieties of patterns and sizes of articles used for one and the same purpose;

(b) to set up standards of quality and dimensions, prepare and promote the adoption of BSI specifications and schedules, revise, alter and amend these from time to time as experience and circumstances may require;

(c) to register, in the name of the Institute, marks of all descriptions, and to approve, affix, and license the affixing of such marks;

(d) to take any other action as may be necessary or desirable to promote the objects of the Institute.

2.114 The Institute is governed by an Executive Board consisting of thirty-five members, who appoint a Director General. Over

1,000 technical committees perform the day-to-day work of developing appropriate standards. When these are agreed, the final standard can be adopted. The Institute is financed from government funds, sales revenue and individual subscriptions.

2.115 Manufacturers who adopt the approved standards are entitled to use the BSI Kite Mark, but must agree to supervision and sample testing.

2.116 The Institute also works closely with the International Organisation for Standardisation and other European organisations which have objects similar to its own.

INDUSTRIAL SAFETY PROTECTIVE EQUIPMENT MANUFACTURERS' ASSOCIATION

2.117 ISPEMA is a trade association which represents the views and activities of manufacturers of protective equipment. It provides technical information to the BSI and represents the industry at international level. It also provides training and educational programmes for industry relating to hazard analysis and the proper selection and usage of personal and environmental protective equipment.

2.118 An associated organisation is the Safety Equipment Distributors Association (SEDA) which comprises distributors of products of ISPEMA members. SEDA feed back information from the user, thus helping in the improvement in the design, comfort, and wearability of the product.

Sources of further information

2.119 For the professional safety adviser, Redgrave's *Health and Safety in Factories* (ed Machin and Fife) contains all the relevant industrial safety law. Equally important is the same author's *Health and Safety at Work*, which contains the Health and Safety at Work etc Act, Offices, Shops and Railway Premises Act, Shop Acts, and so on. Both books are the 'bibles' of safety experts. Also useful in this connection is the *Encyclopaedia of Health and Safety at Work* (ed Goodman) which is in three volumes with an updating service.

2.120 A very useful publication is the *Health and Safety Information Bulletin* (HSIB), a bi-monthly journal published by Industrial

Relations Services, 67 Maygrove Road, London NW6 2EJ. This is essential reading for those who wish to keep themselves fully informed about the latest developments in proposed or actual legislation. It also reports recent decisions relevant to health and safety law, and contains many useful articles on a wide range of subjects.

2.121 Incomes Data Services (IDS), 140 Great Portland Street, London W1 produce a Brief (published fortnightly) which will sometimes report decisions of industrial tribunals on appeals against Prohibition and Improvement Notices, as does Industrial Relations Law Reports (IRLR) published monthly by Industrial Relations Services.

Training films

2.122 A number of organisations make or distribute films on health and safety which are very useful in training sessions, where the message is sometimes more important than the detail. For further information, contact the following organisations:

(a) RoSPA Film Library, Cannon House, Priory Queensway, Birmingham B4 6BS;
(b) Millbank Films Ltd, Thames House North, Millbank, London SW1P 4QG;
(c) Central Film Library, Bronyard Avenue, Acton, London W3 7JB;
(d) HSE Film Catalogue, HSE, Baynards House, 1 Chepstow Place, London W2 4TF.

Environmental studies

2.123 The boundary between occupational health and safety on the one hand, and the wider problems of environmental health and safety on the other is very marginal, and frequently the issues and subject matter are the same. A useful magazine to read is *International Environmental Safety*, which covers the broader implications of safety and health. It is obtainable from Labmate Ltd, Newgate, Sandpit Lane, St Albans, Herts AL4 OBS. Another journal on this topic is *ENDS Report*, published bi-monthly, by Environmental Data Services, Orchard House, 14 Great Smith Street, London SW1.

3 Health and Safety at Work etc Act 1974

The scope of the Act

3.1 The Health and Safety at Work etc Act (HSWA) is based on principles and details which are fundamentally different from other health and safety legislation. These differences are designed to bring about a greater awareness of the problems which surround health and safety matters, a greater involvement of those who are, or who should be, concerned with improvements in occupational health and safety, and a positive movement away from the apathy and indifference which tended to surround the whole subject.

3.2 In the first place, the Act applies to people, not to premises. It covers all employees in all employment situations. We will not be concerned with problems of interpretation as to whether or not certain premises are, or are not, a factory, a shop, an office, etc. The nature or location of the work is irrelevant. At one stroke, the Act brought within the ambit of protective legislation some 7-8,000,000 people (the 'new entrants') who were hitherto not covered by the various statutes in force. Subject only to the exception in respect of domestic employees (see below) every employer needs to know and to carry out his duties under the Act. Further, the Act requires all employers to take account of the fact that persons who are not in his employment may be affected by work activities, and there are additional duties in this regard. Obligations are placed on those who manufacture, design, import or supply articles or substances for use at work to ensure that these can be used in safety and are without risks to health.

3.3 Next, the Act is basically a criminal statute, and does not give rise to any civil liability. No tort action in respect of a breach of statutory duty can be brought, for prevention and punishment, rather than compensation is the key note. Additionally the inspectorates are given new powers of enforcement in order to eliminate or minimise an actual or potential hazard before an incident occurs, rather than take action afterwards.

3.4 Finally, there are some provisions which are designed to bring about a greater personal involvement and individual responsibility so as to actively encourage and promote health and safety, being part of the greater self-regulatory system which the Robens Committee thought to be desirable. Safety policies, safety representatives and safety committees should increase the awareness that the main responsibility for the elimination of accidents and ill-health in employment lies with those who create the dangers.

3.5 How successful the Act has been, or will be in the future, is something we may never know. It is impossible to state how many accidents did not happen as a result of this legislation. Statistics may illustrate trends, and perhaps even lead to some satisfaction, but these are notoriously inconclusive in some respects, and we may never know the real truth. New machinery, new processes, new designs, new substances etc may decrease hazards irrespective of legislative arrangements. If nothing else, the Act should play a positive role in ensuring that technological change does not increase the exposure to risks at work. On the other hand, the increasing consciousness of the problems, the growing acceptance that health and safety is the concern of all – from the most senior person in the organisation to the most junior – the vast increase in training opportunities, and the general streamlining of legal rules, should all have a considerable effect in reducing the annual accident figures progressively each year. There must be a limit, however, to the power of the law to influence human conduct, for a majority of accidents occur in situations where there is no actual breach of a legislative provision, but are regarded as common occurrences or human failings (slipping, tripping, falling, etc), which legal control can only prevent if it can induce a state of mind. A knowledge of the legal framework may make a positive contribution to this.

3.6 The Act is divided into four Parts. Part I will be the subject of this chapter; Part II provides for the continuation of the Employment Medical Advisory Service (see chapter 2); Part III makes amendments to the Building Regulations (and is therefore the main reason for the curious addition of the word 'etc' in the title of the Act), and Part IV contains some miscellaneous provisions. There are also ten Schedules to the Act.

Application of the Act

3.7 The Act applies to all employment in Great Britain (i e England, Wales and Scotland) but not to Northern Ireland. However, by the Health and Safety at Work (Northern Ireland) Order 1978 the provisions of the Act have been repeated so far as Northern Ireland is concerned, with some modifications. The Act does not apply to the Isle of Man or to the Channel Islands, but does apply to the Isles of Scilly.

3.8 By the Health and Safety at Work (Application outside Great Britain) Order 1977 the provisions of the Act have been extended to cover persons working on offshore installations and pipeline work within British territorial waters or the United Kingdom sector of the continental shelf, as well as to certain diving and construction work carried on within territorial waters. These provisions are in addition to the Mineral Workings (Offshore Installations) Act 1971 and the Regulations made thereunder and the Petroleum and Submarine Pipelines Act 1975. Although HSC will be responsible for the health and safety of such offshore workers, the Secretary of State for Energy retains responsibility for the structural safety of offshore installations and pipelines.

APPLICATION TO THE CROWN

3.9 The provisions of Part I apply to the Crown with the exception of sections 21–25 and 33–42. This means that it is not possible to issue Improvement or Prohibition Notices against the Crown or in respect of Crown premises (including, for example, National Health Service hospitals). However, HSE has taken to issuing Crown Notices (see para 3.77) which have a moral effect. Also, it is not possible to prosecute the Crown for any offence committed (based on the theory that the Queen cannot be prosecuted in her own courts), but section 48(2) provides that sections 33–42 (i e the criminal sanctions sections) shall apply to persons in the service of the Crown as they apply to other persons. This means that an individual Crown employee who commits an offence (e g under sections 7, 8, 37 etc) can be prosecuted if necessary.

DOMESTIC EMPLOYEES

3.10 Section 51 provides that none of the statutory provisions shall apply in relation to a person by reason only that he employs another, or is himself employed, as a domestic servant in a private household.

CIVIL LIABILITY (SECTION 47)

3.11 As we have already noted, HSWA is essentially a criminal statute enforced by criminal sanctions. Section 47 specifically provided that nothing in Part I (which contains the relevant provisions so far as this book is concerned) shall be construed as conferring any right of action in any civil proceedings in respect of a failure to comply with any duty imposed by sections 2–7 or a contravention of section 8. Thus no civil action under HSWA can be brought for a breach of statutory duty (compare, for example, the situation with regard to the Factories Act etc, see chapter 4) as a result of an accident. However, if a prosecution is brought under the Act, and is successful, it would appear that under the provisions of section 11 of the Civil Evidence Act 1968 the fact of prosecution may be raised and pleaded in an action for damages caused by negligence, leaving the defendant with the burden of proving that he was not negligent. Additionally, section 47(2) provides that any breach of duty imposed by health and safety Regulations shall be actionable in a civil claim except in so far as the Regulations provide otherwise. HSWA does not alter the present rights at common law (see chapter 9), but does not add to them except as stated.

The general duties (section 2(1))

3.12 It shall be the duty of every employer to ensure, so far as is reasonably practicable, the health, safety and welfare at work of all his employees. This is the prime duty under the Act, in respect of which all the other subsequent duties imposed by section 2 are more detailed. As we have seen (see chapter 1), what is reasonably practicable is a question of fact and evidence in each case.

3.13 The duty only extends to employees while they are in the course of their employment; there is no obligation on an employer to concern himself with health, safety or welfare matters which arise outside work, although this is being done increasingly as part of advances in personnel policies.

3.14 The phrase 'health, safety and welfare' is not defined. Clearly, health includes mental as well as physical health, and safety refers to the absence of any foreseeable injury. Welfare, on the other hand, is a somewhat elusive concept. Schedule 3 of HSWA provides that Regulations may be made for 'Securing the provision of specified welfare facilities for persons at work, including, in particular, such things as adequate water supply, sanitary conveniences, washing and bathing facilities, ambulance and first aid arrangements, cloakroom accommodation, sitting facilities and refreshment facilities.' It is clear that this list is not an exhaustive definition of welfare, and indeed, some of them may well be regarded as health matters. Social clubs may be generally regarded as being part of a firm's welfare facilities, but since employees do not use them in the course of their employment, they are not within the scope of the Act.

Particular duties (section 2(2))

3.15 The above general duty is particularised by five specific duties which are placed on the employer, which spell out the general duty in detail.

3.16 (a) To provide and maintain plant and systems of work that are, so far as is reasonably practicable, safe and without risks to health. An employer 'provides' when the plant etc is in a place where it can be easily come by, or when he gives clear directions as to where it can be obtained (*Norris v Syndic Manufacturing Co*). Thus in *Woods v Durable Suites* the employee was working with a synthetic glue, which could cause dermatitis unless certain precautions were taken. The employers instructed the employee in the proper procedure to be followed, and provided washing facilities and a barrier cream, but the employee did not take the necessary precautions and contracted dermatitis. It was held that the employers had fulfilled their common law duty of care, for they had provided the necessary precautions. They were under no duty to compel the employee to use the barrier cream or to stand over him to ensure that he used the washing facilities.

3.17 The duty to provide may be observed even though the employee does not use that which is provided, although in certain cases it has been held that the employers' common law duty may be something more than the passive one of providing, and includes the more active duty of encouraging, persuading and even insisting on the use of the precautions (see chapter 9). On other occasions the law may well go further, and place an obligation on the employee *to use* the precautions supplied, e g under the Protection of Eyes Regulations 1974, see chapter 4. But if the employer fails to provide the necessary equipment, or provides defective equipment, he will be in breach of his duty. In *Lovett v Blundells and Crompton & Co* the employee erected a makeshift staging in order to do his work. This collapsed, and he was injured. It was held that the employer had not provided adequate equipment for the work to be done in safety.

3.18 Once having provided the necessary plant, etc, the employer must ensure that it is maintained in proper working order, and in a condition which makes it fit for use. Maintenance is a matter of forethought and foresight. There must be a proper system of regular inspection, with the reporting of defects to a responsible person. In *Barkway v South Wales Transport Co* a coach crashed owing to a burst tyre, and the plaintiff's husband was killed. Although the defendants could show a system of testing and inspecting tyres, they did not require their drivers to report incidents which could produce impact fractures, and were thus held liable for negligence.

3.19 Maintenance also requires the rectification of known defects, either by repairs or replacement, as necessary. In *Taylor v Rover Car Co* the employee was injured when a splinter of steel flew from the top of a chisel he was using. The chisel had previously been used by a leading hand on the production line, who had himself been injured when a splinter had flown off, but this incident was not reported. It was held that the employers were liable, for they should have had a system whereby defective tools were reported and withdrawn from circulation and replaced by ones which were not defective.

3.20 Routine maintenance, as well as revealing defects, can prolong the life of the plant or machinery, etc, thus creating a cost benefit for the employer as well as making it safe for the user. It is further suggested that manufacturers of articles for use at work should include maintenance schedules along with the articles they sell, as being part of 'the conditions necessary to ensure that when put to use it will be safe and without risks to health' (see section 6, below).

3.21 Section 53(1) states that the term 'plant' includes any machinery, equipment or appliance. In accordance with the *ejusdem generis* rule of interpretation, it is unlikely that 'plant' can thus mean premises of any kind (contrary to popular usage of the word). However, the definition is not exhaustive, and there is no guidance as to what else is included.

3.22 Systems of work was defined by Lord Greene in *Speed v Thomas Swift & Co Ltd*, who said, 'It may be the physical layout of the job – the setting of the stage, so to speak – the sequence in which the work is to be carried out, the provision in proper cases of warnings and notices, and the issue of special instructions'. To this, we may add the provision of safety equipment and the taking of adequate safety precautions. Whatever system of work is adopted, the employer must ensure, so far as is reasonably practicable, that it is a safe one.

3.23 (b) Arrangements must be made for ensuring, so far as is reasonably practicable, safety and the absence of risks to health in connection with the use, handling, storage and transport of articles and substances. Thus in appropriate cases, protective clothing, proper equipment and tools, etc must be provided. The handling must be organised in a safe manner. Excessive weights should be considered, contamination should be guarded against, dangerous parts should be covered etc. Storage facilities must be adequate and safe, e g proper racks provided, fork lift truck drivers must be instructed in proper stacking techniques. The transportation must be done in a safe manner: loads must be properly tied down, with an even distribution of weight, goods must be packed properly so as to be safe when being transported. In *Page v Freight Hire (Tank Haulage) Ltd* the applicant was a twenty-three year old female heavy goods vehicle driver employed on a causal basis to drive a tanker which contained dimenthylformamide (DMF). The employers were then informed that there was a danger to women of child-bearing age if they came into contact with this chemical, and so her employers refused to allow her to drive the tanker. She claimed that this constituted unlawful sex discrimination contrary to section 6(2)(b) of the Sex Discrimination Act 1975. However, section 51(1) of that Act excludes its operation when it is necessary to do something in order to comply with a requirement of an Act of Parliament passed before 1975. The EAT were able to point to the fact that the employers were under certain obligations to comply with section 2(2) of HSWA (passed, of course, in 1974), and thus there was no unlawful discrimination. The interests of safety clearly required that she should not be exposed to the handling of a substance which was potentially hazardous, and which had a possible embryotoxic effect.

3.24 (c) The provision of such information, instruction, training and supervision as is necessary to ensure, so far as is reasonably practicable, the health and safety at work of all his employees.

3.25 Thus, information must be given to employees about the hazards involved in the work, and the precautions to be taken to avoid them. Since the employer will, in most cases, be in a better position to know of those hazards, he must provide the information, not wait for the employees to request it.

3.26 The information which is provided must be accurate and meaningful. In *Vacwell Engineering Ltd v BDH Chemicals Ltd* the plaintiff purchased a quantity of boron tribrimide from the defendant. The chemical was delivered in ampoules which were labelled 'Harmful vapour'. The chemical was poured on to some water, and an explosion resulted, causing damage to the plaintiff's premises. It was held that the defendants were liable for negligently labelling a dangerous substance. The information given was misleading, and did not accurately describe the hazard.

3.27 Proper and clear instructions must be given as to what is to be done, and what must not be done, for workers performing routine tasks are frequently heedless of their own safety. Greater care must be taken when dealing with employees whose command of English is weak, so that they understand clearly the nature of the dangers and the precautions to be taken. Young and inexperienced workers must be given clear instructions.

3.28 The employer's duty to provide information and instruction to ensure the health and safety of his own employees includes a duty to provide such information and instruction to the employees of a subcontractor where this is necessary to ensure the health and safety of the employees. In *R v Swan Hunter Shipbuilders Ltd*, the employers distributed a booklet to their employees giving practical rules for the safety of users of oxygen equipment, in particular warning of the dangers of oxygen enrichment in confined spaces. The booklet was not distributed to employees of a subcontractor except on request. A fire broke out on a ship which the appellants were building, and because an employee of a subcontractor left an oxygen hose in the deck, the fire became intense because that part of the ship, which was badly ventilated, became oxygen enriched. As a result, eight workmen were killed. The appellants were prosecuted on indictment for a breach of section 2(2)(a) (failure to provide a safe system of work), section 2(2)(c) (failure to provide information and instruction) and section 3(1) (duty to non-employees, see below). The appellants were convicted in the Crown Court, and an appeal was dismissed by the Court of Appeal. If the provision of a safe system of work for the benefit of an employer's own employees involved the provision of information and instruction as to potential dangers to persons other than his own employees, then the employer was under a duty to provide such information and .instruction. In the circumstances, is was reasonably practicable to do so.

3.29 Training in safe working practices should be undertaken on a regular basis; where special courses are available, employees should be required, not merely encouraged, to go on them (e g for fork lift truck drivers, or when handling heavy weights). Where appropriate, 'in-house' training can be given. The employee is under a legal duty to co-operate with the employer (section 7(b), below), and hence a refusal to go on an appropriate training course, as well as being a possible offence under the Act, may well be grounds for fair dismissal (see chapter 10). Sometimes, Regulations will specify the actual training to be given (e g Abrasive Wheels Regulations 1970), including the syllabus, but otherwise the training may be carried out by the employer, or an independent body. As long as it is adequate, the requirements of HSWA will be met.

3.30 A suitable and satisfactory system of supervision must be provided, with properly trained and competent supervisors, who have authority to ensure that safety precautions are implemented, safety equipment used, and safe systems followed. Young and inexperienced employees in particular must be properly supervised.

3.31 (d) So far as is reasonably practicable as regards any place of work under the employer's control, he must ensure its maintenance in a condition that is safe and without risks to health, and the provision and maintenance of means of access and exit that are safe and without such risks. Thus premises must be safe and maintained safe. Obstacles must be removed, dangerous wiring replaced, defective floors and stairs repaired, roads, pavements, pathways, doors, etc must all be safe.

3.32 (e) The provision and maintenance of a working environment that is, so far as is reasonably practicable, safe, without risks to health, and adequate as regards facilities and arrangements for the welfare of employees at work. Thus the employer must pay proper regard for systems of noise control, eliminate noxious fumes and dust, lighting must not be excessive or inadequate. Welfare arrangements and facilities must be adequate, e g toilet accommodation, washing facilities, cloakroom arrangements, etc.

3.33 It will be recognised that the above statutory duties bear a strong resemblance to those duties owed by an employer to an employee at common law (see chapter 9) spelt out in greater detail. Under section 2 of HSWA, these duties may only be enforced by criminal sanctions or by the use of an enforcement notice (see para 3.77).

Safety policies (section 2(3))

3.34 Except for employers who employ less than five employees, every employer shall prepare, and as often as may be appropriate revise, a written statement of his general policy with respect to

(a) the health and safety at work of his employees, and
(b) the organisation and arrangements in force for the time being for carrying out that policy,

and bring the statement and any revision of it to the notice of all his employees.

3.35 The drawing up of the safety policy is the beginning of the commitment of the employer to safety and health at work. There is no standard policy, no precedent which can be adopted. Each employer must work it out for himself, bearing in mind the nature of the hazards involved, and the precautions and protections needed. General advice on the drawing up of safety policies may be sought from a number of sources (employers' associations, safety organisations, etc) but the responsibility is placed fairly and squarely on the shoulders of each employer. Studies have revealed that whilst some policy statements lay down a general commitment, they lack details of 'the organisation and arrangements in force' for the carrying out of the policy. Moreover, there is a need to constantly revise the policy statement when appropriate.

3.36 The Act requires the statement to be brought to the notice of all employees. This is not done merely by affixing a policy statement to a notice board. Some employers issue the statement as a paper communication, but this, whilst conforming with the law, is not necessarily good practice, for sheets of paper are frequently lost or destroyed. Perhaps the best method is by introducing the policy statement on an induction course, and publishing it together with the works rules or information handbook which is given to employees. Particular attention should be paid to those employees whose command of English is limited, and steps must be taken to draw attention to the policy in a language they understand.

3.37 A safety policy can be drawn up in the following manner:

i *General statement*

3.38 This will specify the commitment of the employer, at a standard at least as high as that required by law. It should make clear that management considers it to be a binding commitment, and that safety will rank as a prominent and permanent feature of all activities. The objectives should be spelt out, e g to reduce and eliminate accidents, to achieve a safe and healthy working environment, and so forth.

ii *Organisation*

3.39 The distribution of responsibility should be detailed, starting with the management board, through the different levels of management, supervision, safety officer and safety representatives, medical personnel, and ending with the responsibilities of employees. Where there are a number of sites or departments, responsibilities should be fixed as appropriate. It is probably good practice to identify the person responsible in each case, either by name or by position. The lines of communication for dealing with grievances, complaints or suggestions about health and safety matters should be stated.

iii *Arrangements in force*

3.40 The existence of the arrangements must be stated in relation to the objectives to be achieved in each case. For example, on health, state details of the first aid facilities, fire precautions, medical arrangements, etc. On safety, specify training, supervision, safety equipment, safety precautions, safety rules, maintenance practices, etc. On welfare, specify washing facilities provided, requirements relating to ventilation, heating, lighting, etc. Stress the need for all employees to be involved in good housekeeping, to co-operate with management, and to report any defects or potential hazards.

iv *Review*

3.41 Safety policies cannot be adequately reviewed unless there is periodic monitoring. This may well be one of the functions of the safety committee, but the prime responsibility must always rest with senior management. The statement should be dated and signed by the senior person in the organisation so that employees will recognise that it is an authoritative document and will note the ongoing commitment.

3.42 Safety policies should be seen to work. They should not be a mere formality to satisfy the curiosity of visiting inspectors.

General duties owed to others (section 3)

3.43 Every employer is under a duty to conduct his undertaking in such a way as to ensure, so far as is reasonably practicable, that persons not in his employment who may be affected thereby are not exposed to risks to their health and safety. A similar duty is imposed on self-employed persons, in respect of himself and other persons not being his employees (section 3(2)). The Health and Safety (Genetic Manipulation) Regulations 1978 provide that the reference to a self-employed person shall include a reference to any person who is not an employer or employee, in relation to any activity involving genetic manipulation. This would include, for example, a research student.

3.44 Section 3 is designed to give protection to the general public, to ensure that they are not at risk from industrial hazards, etc. Thus, it would be an offence under the Act (irrespective of any other heading of legal liability) if a construction firm were to permit an explosion to take place which causes windows to break in nearby houses, or for a firm to permit the seepage of a poisonous chemical into a private water supply. Visitors who come on to the employer's premises, subcontractors who come to work there, students who are on a University campus, etc, are all within the class of persons who may be affected by the way the undertaking is being carried on. Whether such people have a right of civil action is irrelevant to the criminal liability of the employer in respect of a breach of the duty.

3.45 The duty imposed by section 3(1) is wide enough to include the duty to provide information and instruction to persons who are not in the employer's employment (*R v Swan Hunter Shipbuilders Ltd*).

3.46 The words 'conduct his undertaking' do not appear to apply to the effects which a deleterious product may have on an ultimate consumer, where the ordinary law of negligence (as in *Donoghue v Stevenson*) will apply (or product liability, if it is ever brought into force in this country).

3.47 In such cases as may be prescribed, and in the circumstances and prescribed manner, it shall be the duty of every employer and every self-employed person to give to persons who are not his employees, but who may be affected by the way he conducts his undertaking, the prescribed information about such aspects of the way he conducts his undertaking as might affect their health or safety (section 3(3)). This subsection is clearly designed to ensure that persons living near to some hazardous operation have some form of advance notification of what they may expect if something goes wrong, and the necessary action they should take. To date, no Regulations requiring the disclosure of any such information have been made.

General duties of controllers of premises (section 4)

3.48 This section imposes duties on people who have the control over non-domestic premises, or of the means of access thereto or exit therefrom, or of any plant or substances therein, which are used by persons who are not employees as a place of work or as a place where they may use plant or substances provided there for their use. The duty is to take such measures as it is reasonable for a person in his position to take to ensure, so far as is reasonably practicable, that the premises, the means of access and exit, and any plant or substance in or provided for use there, is or are safe and without risks to health.

3.49 Section 4 is the criminal counterpart of the civil liability contained in the Occupiers' Liability Act 1957 (and the Occupiers' Liability (Scotland) Act 1960) and would apply in those cases where an employee is using premises which are not controlled by the employer, e g a visiting window cleaner. However, the effect of the section is somewhat wider than that. Thus a coin-operated laundrette would be covered, even though no-one was employed there, for a customer would be using plant or substances provided for use. Universities would have a duty under this section to ensure that their premises (libraries etc) are safe; schools have duties towards their pupils using laboratories, etc.

3.50 Further, a person who, by virtue of any contract or tenancy, has an obligation to maintain or repair any premises used by others as a place of work or as a place where they may use plant or substances, or to maintain the means of access or exit, or to ensure the safety or absence of risks to health arising from the use of such plant or substances, will be the person upon whom the above duty will lie. For example, a maintenance contractor who is responsible for the maintenance of plant, or a specialist adviser who has to deal with the control of a dangerous substance, will have those duties imposed by this section to persons other than his employees.

3.51 However, in each of the above cases, the premises must be carried on by the person in control by way of a trade, business or other undertaking (whether for profit or not). The question may well arise as to whether the word 'undertaking' is to be construed *ejusdem generis* with 'trade or business', as implying some form of commercial activity.

Duty to prevent pollution (section 5)

3.52 Every person having control of any premises of a class prescribed shall use the best practicable means for preventing the emission into the atmosphere from the premises of noxious or offensive substances, and for rendering harmless and inoffensive such substances as may be emitted. The 'means' to be used include a reference to the manner in which the plant is used, the supervision of any operation involving the emission of the noxious or offensive substance, as well as any device used to prevent them entering the atmosphere, or for rendering them harmless. To date, no class of premises have been prescribed for the purpose of this section.

3.53 The 'best practicable means' is the highest standard short of absolute liability. An indication of what this will currently be is contained in Guidance Notes which are issued by the Alkali and Clean Air Inspectorate, and it is therefore necessary to ensure that the latest technological and scientific developments are noted and implemented.

General duties of designers, manufacturers, importers and suppliers (section 6)

3.54 Section 6 lays down four duties on any person who designs, manufactures, imports or supplies any article for use at work, and on any person who manufactures, imports or supplies any substance for use at work. Before we consider the nature of these duties, we must first ascertain the scope of the above words.

3.55 'An article for use at work' is defined as being 'any plant designed for use or operation (whether exclusively or not) and any article designed for use as a component in any such plant'. The word 'plant' includes 'any machinery, equipment or appliance'. A 'substance' is 'any natural or artificial substance, whether in solid or liquid form, or in the form of gas or vapour' (section 53). 'Work' in this connection, means at work as an employee or as a self-employed person. Thus a sale to a 'do-it-yourself' enthusiast is not within section 6, for although he may be working, he is not at work. Consumer sales are also not within the Act, but since section 53 states that the article need not be exclusively designed for use at work, an item which is capable of being used at work is within the scope of the section even though it is also capable of being a consumer item. It may be that a test will emerge which asks if it was reasonably foreseeable that an employer would purchase the item for use at work.

3.56 The article or substance must be 'for use'. Thus if it is part of the stock-in-trade, or is purchased for resale purposes, the section does not apply. Nor does it apply to goods which are manufactured etc for export, for the Act does not apply extra-territorially.

3.57 'Supply' in this connection means the supply of an article or substances by way of sale, lease, hire or hire purchase, whether as a principle or as an agent for another. However, section 6(9) recognises the commercial nature of hire-purchase agreements, conditional sales agreements, and credit-sales agreements, and draws a distinction between the ostensible supplier and the effective supplier. The ostensible supplier is in reality merely financing the transaction even though in the course of the transaction he may become the legal owner of the goods in question. The liability under section 6 is on the effective supplier, i e the manufacturer, etc, who sells to the finance company, who then sells to the customer. A similar provision is to be found in the Health and Safety (Leasing Arrangements) Regulations 1980, whereby if the ostensible supplier is merely acting as a financier for a leasing arrangement made between the effective supplier and the customer, then, subject to certain conditions, the effective supplier and not the ostensible supplier will have the duties of section 6 imposed on him.

3.58 It will be noted that in respect of articles, the designer has certain duties, as well as manufacturers, importers and suppliers. It is not clear whether this means the person who actually designs the product, or the employer of the designer. It is likely that

HSE policy will be to leave the employee to be prosecuted under section 7, and take action against the employer of the designer under section 6. This is because subsection (7) of section 6 reminds us that the duties only apply 'to things done in the course of a trade, business or undertaking carried on by him' and presumably it is the employer of the designer who is carrying on the trade business or undertaking. If the employer had no reason to suspect that the designer was incompetent, or had made a faulty design, presumably he could rely on the defence that he took all steps, etc which were reasonably practicable.

3.59 The four duties under section 6 are as follows:

3.60 (a) To ensure, so far as is reasonably practicable, that the article is so designed and constructed as to be safe, or the substance is safe, and without risks to health when properly used (section 6(1)(a) and (4)(a)). The expression 'safe, and 'without risks to health when properly used' has given rise to a certain amount of discussion as to its meaning. One school of thought takes the view that it means that it must be supplied in a safe condition; 'safe' is thus equated with 'not dangerous'. The other view is that it can be supplied in an unsafe condition, but providing it is accompanied by adequate information about the matters which need to be done before it is put into use, then it is still 'safe' within the statutory meaning. On the first view, a machine is not safe if it is supplied without guards; on the second view, it is safe if the users' attention has been drawn to the need to fix suitable guards. The second – and narrower – interpretation is further supported by subsection (10), which states that an article or substance is not to be regarded as properly used where it is used without regard to any relevant information or advice relating to its use which has been made available by the person by whom it was designed, manufactured imported or supplied. For example, in a Scottish case, the supplier of a mincing machine was prosecuted for supplying a machine which had an unguarded opening into which the meat was fed, and which permitted the operator to insert his hands. The defendant argued that a wooden plunger was supplied with the machine, and providing this was used there was no danger. The sheriff accepted that though the machine was intrinsically unsafe, it was 'safe . . .when properly used', and the summons was dismissed. A similar view appears to have been taken in other cases, and though they clearly do not have the force of binding precedents, it may be of significance that the prosecution did not consider taking the cases to appeal.

3.61 In a Guidance Note on section 6, HSE takes the view that 'If the user makes an unusual or unexpected use of the product, or chooses to disregard the information (otherwise than in some small and/or irrelevant particular), by deliberately overloading a machine, failing to install a machine correctly for noise control purposes, or failing to maintain equipment, for example, then it will be considered that there has not been a proper use and the manufacturer etc will not, under those circumstances, be held responsible under section 6.'

3.62 (b) To carry out or arrange for the carrying out of such testing and examination as may be necessary for the performance of the duty imposed on him by the preceding paragraph (section 6(1)(b) and (4)(b)). However, a person is not required to repeat any testing or examination which has been carried out by some other person in so far as it is reasonable for him to rely on the results thereof (section 6(6)).

3.63 (c) To take such steps as are necessary to ensure that there will be available in connection with the use of the article or substance at work, adequate information about the use for which the article has been designed and tested (section 6(1)(c)) and about any relevant tests which have been carried out on or in connection with the substance (section 6(4)(c)) and about any conditions necessary to ensure that it will be safe and without risks to health when used. Again, the meaning of the section is not clear. Does this imply that the supplier etc must also supply the information, or is his duty confined to having the information available for anyone who may request it? The better view appears to be that the information is available only when the attention of the user has been drawn to its existence, for example, by the necessary documentation or by a service manual which accompanies the product. Whether this interpretation is strictly in accordance with the literal meaning of the Act remains to be decided. However, reference may again be made to subsection (10) which will excuse the supplier etc if the product has been used without regard to any information which has been made available.

3.64 (d) A designer or manufacturer of an article, or the manufacturer of a substance, is under a duty to carry out or to arrange for the carrying out of any necessary research with a view to the discovery and, so far as is reasonably practicable, the elimination or minimisation of any risks to health or safety to which the design of the article, or the substance, may give rise (section 6(2) and (5)). Again, there is no need to repeat any research which has been carried out by some other person, in so far as it is reasonable to rely on the results thereof (section 6(6)). This is the first occasion in English law where there is a legal obligation to carry out research.

3.65 The above duties only extend to a person in respect of things done in the course of a trade, business or other undertaking carried on by him (whether for profit or not) and to matters within his control. This appears to exclude employees from the scope of section 6 (though not, of course, from section 7). Whether the matters are within the control of a person is a question of fact. If he has the right of control, but fails or refuses to exercise it, the matters are still within his control.

Duties of erectors and installers (section 6(3))

3.66 A person who erects or instals any article for use at work in any premises where that article is to be used by persons at work shall ensure, so far as is reasonably practicable, that nothing about the way in which it has been erected or installed makes it unsafe or a risk to health when properly used.

Indemnity clauses (section 6(8))

3.67 Where a person designs, manufactures imports or supplies an article to or for another person on the basis of a written undertaking by that other person to take specific steps sufficient to ensure, so far as is reasonably practicable, that the article will be safe and without risks to health when properly used, the undertaking will have the effect of relieving the first-mentioned person from the duty of ensuring that it is designed and constructed so as to be safe, to such an extent as is reasonable having regard to the terms of the undertaking. Thus if a person supplies secondhand machinery to another on the basis of that other's assurance that he will ensure that it is properly serviced and examined before being put to use, or where a manufacturer supplies machinery made specifically to certain specifications or to a certain design, the supplier or manufacturer may be relieved from his legal responsibility under section 6(1)(a). However, there must be a written assurance, which implies a specific commitment in the instant case, not a general standard commitment. Further, the exclusion is only from the liability under section 6(1)(a), not from the liability imposed by section 6(1)(b) or 6(1)(c) or 6(2). Further, the terms of the undertaking may be looked at to discover the extent to which the designer etc is to be absolved. HSE appear to take the view that this may lead to a partial relief, depending on the terms of the undertaking, but this may be misleading. A breach of section 6 is a criminal offence; either an offence has been committed or it has not. There is no such thing as 'partially guilty'. Mitigating circumstances can only arise if an offence has been committed, and the extent of the mitigation will be for the court to determine.

The effect of section 6

3.68 The exact scope and meaning of section 6 has still to receive authoritative judicial interpretation, for although there have been some prosecutions, there appears to be a marked reluctance on either side to challenge these findings in the High Court. The purpose it is try to ensure that acceptable levels of health and safety are built into articles and substances at the design and manufacturing stage, whether by way of compliance with recognised standards (e g BSI) or HSE Guidance Notes or other acceptable test. But the fact that the manufacturer etc may be in breach of his duty under section 6 is irrelevant to the employer's liability under the general law to take reasonable care to ensure the health and safety of his employees, or the absolute duty to fence dangerous machinery (section 14 of FA). A modern practice is for purchasers of products for use at work to make the contract conditional upon compliance by the supplier with section 6 of HSWA, thus leaving it open for the purchaser to reject the product. It will be recalled that nothing in HSWA gives rise to any civil liability, but it would be an interesting argument if the purchaser rejected a product on the grounds that a failure to comply with the requirements of section 6 rendered the product not of merchantable quality as required by the Sale of Goods Act 1979.

General duties of employees (section 7)

3.69 Two main duties are placed on employees.

3.70 (a) To take reasonable care for the health and safety of himself and of others who may be affected by his acts or omissions at work. Thus an employee who refuses to wear safety equipment or use safety precautions is liable to be prosecuted under this section. Further, if through his carelessness or negligence someone else is injured, he could again be prosecuted. Thus an employee who is prone to horseplay or skylarks around, with the result that he or another is injured, commits an offence; a supervisor who encourages an employee to take an unsafe short cut or to remove effective guards may equally be guilty under this section.

3.71 (b) As regards any duty or requirement imposed on his employer or other person by or under any of the relevant statutory provisions, to co-operate with him in so far as is necessary to enable that duty or requirement to be performed or complied with. This duty to co-operate is potentially very wide. An employee who announces that he intends to refuse to wear a safety belt, or use a safety precaution provided by his employer in pursuance of the latter's duty under section 2 is failing to co-operate with the employer, and thus a prosecution may again succeed. One wonders what the position would be if safety inspectors went on strike!

Interference or misuse (section 8)

3.72 No person shall intentionally or recklessly interfere with or misuse anything provided in the interests of health, safety or welfare in pursuance of any of the relevant statutory provisions.

3.73 This obligation is again wider than the corresponding provision in the Factories Act, which referred to wilful conduct, in the sense of being deliberate or perverse. Intentional or reckless conduct does not need to be perverse.

Duty not to charge (section 9)

3.74 No employer shall levy or permit to be levied on any employee of his any charge in respect of anything done or provided in pursuance of any specific requirement of the relevant statutory provisions. Thus, if an employer is obliged by law to provide specific safety precautions or safety equipment (e g appropriate clothing) he may not charge for them. If, on the other hand, he provides them in pursuance of his common law duty or his general duty under section 2(1), and charges for them, he does not commit an offence under section 9. For example, safety boots may be provided and charged for unless there is a statutory requirement to provide them. In practice, employers will issue certain items free of charge or at cost price if they are capable of being used outside the employment. There appears to be a growing assumption that all safety equipment should be provided without charge, a view which stems from the interpretation of section 2(1) of HSWA (see *Associated Dairies v Hartley*, para 3.97) or as a result of there being an implied term in the contract of employment to this effect (see *British Aircraft Corporation v Austin*, chapter 10). However, this may be so only in so far as it is reasonable or reasonably practicable to provide such equipment.

3.75 A list of the specific statutory requirements can be found in chapter 6.

Enforcement of the Act

3.76 A breach of the Act or of health and safety Regulations can be dealt with in two ways. First, there are the new powers given to the inspectors to issue enforcement notices or to seize and destroy. Second, a prosecution may take place in respect of the commission of a criminal offence.

Enforcement Notices (sections 21–24)

3.77 There are two types of Enforcement Notices which may be issued. These are (1) Improvement Notice, and (2) Prohibition Notice (which may be immediate or deferred). In addition, the inspectorate have taken to issuing Crown Notices in respect of premises belonging to (or under the control of) the Crown (e g National Health Service hospitals), but although there is no legal basis for such Notices (it will be recalled that sections 21–24 do not bind the Crown) they have a moral and persuasive effect. A failure to comply with a Crown Notice would lead to an approach by HSE to the Government Department concerned, and the Government has announced that in such circumstances the necessary action would be taken to ensure compliance. Moreover, a copy of the Crown Notice will be given to the representatives of the employees, thus drawing attention to the hazard.

3.78 Enforcement Notices may be issued by HSE inspectors, Environmental Health Officers of a local authority, and the Railway Inspectorate in respect of railway premises, all of whom must act within the powers contained in the instrument of their appointment. However, it must be shown that the premises concerned are within the scope of the relevant legislation. In *Dicker & Sons v Hilton* a Notice was served requiring the appellant to comply with section 36 of the Factories Act which lays down that air receivers shall be cleaned and examined by a competent person every twenty-six months (see chapter 4). The Notice was cancelled on appeal when the industrial tribunal learned that the appellant ran a one-man business. Since his premises were not a factory (which is defined as being premises where persons are employed) the relevant statutory provision did not apply.

Improvement Notice (section 21)

3.79 If an inspector is of the opinion that a person

(a) is contravening (or failing to comply with) a relevant statutory provision, or
(b) has contravened (or has failed to comply with) one or more provisions in circumstances that make it likely that the contravention will continue or be repeated,

then he may serve an Improvement Notice, which must state

(a) that he is of that opinion,
(b) the provisions in question,
(c) particulars of the reasons for his opinion,

and requiring that person to remedy the contravention or matters occasioning the contravention within such period as the Notice may specify, but not earlier than twenty-one days after the Notice has been served (which is the period in which the person affected may lodge an appeal against the Notice).

3.80 In other words, if there is a statutory requirement that a certain thing shall be done (or not done) an inspector may serve a Notice requiring the thing be done (or not done) any time after twenty-one days. But the fact that a period of grace is permitted does not absolve the person concerned from any criminal or civil liability in respect of anything which may happen prior to the Notice taking effect.

3.81 An inspector may (but is not bound to) attach a schedule of the remedial steps to be taken (section 24). If he does, and it is unclear or vague, this will not affect the validity of the Notice (*Chrysler (UK) Ltd v McCarthy*) but the industrial tribunal may clarify or alter the schedule.

3.82 If the requirement of the statutory provision is absolute, then there can be no defence in the case of a breach (*Ranson v John Baird*) but if the statutory requirement is to do that which is reasonably practicable, the industrial tribunal may exercise its own judgment in accordance with the circumstances of the case when hearing an appeal. Thus in *Roadline (UK) Ltd v Mainwaring*, an Improvement Notice required an employer to provide heating in a transit shed. The industrial tribunal thought that the cost of doing so was excessive in relation to the marginal improvement which would result.

Prohibition Notice (section 22)

3.83 If an inspector is of the opinion that activities are about to be or are being carried on in relation to which any of the relevant statutory provisions apply, and which may or will involve a risk of serious personal injury, he may serve a Prohibition Notice. This will

(a) state that the inspector is of the opinion;
(b) specify the matter which in his opinion is giving or will give rise to the risk of serious personal injury;
(c) if the matter also involves a contravention of a relevant statutory provision, he will state the statutory provision, and give particulars of the reason as to why he is of that opinion; and
(d) direct that the activities to which the Notice relates shall not be carried on by or under the control of the person on whom the Notice has been served unless the matters specified in the Notice (and any associated contraventions of statutory provisions) have been remedied.

3.84 If the inspector is of the opinion that the risk of serious personal injury is imminent, the Notice will take immediate effect. If it is not imminent, the Notice (which will be a deferred Prohibition Notice) will take effect at the end of the period specified in the Notice. Again, the inspector may, but is not bound to, attach a schedule of the remedial steps to be taken. It will be noted that to issue a Prohibition Notice, the inspector need only be satisfied that the activities complained of give rise to a risk of serious personal injury, and there is no need for there to be a breach of a relevant statutory provision (*Roberts v Day*). However, if he has little information on which to form such an opinion, the Notice may be cancelled by the industrial tribunal (*Bressingham Steam Preservation Co Ltd v Sincock*).

Supplementary provisions (section 23)

3.85 As already noted, an Improvement and Prohibition Notice may (but need not) include directions as to the measures to be taken to remedy any contravention, and these may be framed by reference to any Approved Code of Practice or may give a choice as to different ways of taking remedial action. However, if the Improvement Notice refers to a building, or a matter connected with a building, the Notice may not direct that measures shall be taken which are more onerous than the requirements of any building regulations which are applied in respect of new buildings. If the Notice refers to the taking of measures affecting the means of escape in the case of fire, the inspector must first consult the Fire Authority.

3.86 In the case of an Improvement Notice or a deferred Prohibition Notice, these may be withdrawn at any time by the inspector before the date on which they are to take effect, and also the period for compliance may be extended by him provided an appeal is not pending.

3.87 Once the matter which has been the subject of an Improvement or Prohibition Notice has been attended to, or the person to whom it is addressed has complied with any requirement contained therein, the activity may be carried on without any further need to contact the inspector, although prudence may well advise such a course in order to ensure that he is satisfied with the rectification. This is particularly important in view of the fact that a failure to comply with the requirements of an Enforcement Notice exposes the offender to potentially severe punishment.

Appeals against Enforcement Notices (section 24)

3.88 A person on whom a Prohibition or Improvement Notice has been served may appeal to an industrial tribunal within twenty-one days of its receipt, and the tribunal may confirm or cancel the Notice. If it is confirmed, this may be done in its original form, or with such modifications as the industrial tribunal thinks fit. An appeal may be made on a point of law or of fact.

Procedure for appeals

3.89 The Industrial Tribunal (Improvement and Prohibition Notices Appeals) Regulations 1974 lay down the procedure to be followed for the making of an appeal against the decision of an inspector to issue a Notice. The appellant shall send a Notice of Appeal to the Central Office of Industrial Tribunals (in London or Glasgow, as appropriate):

(a) stating his name and address for service of documents;
(b) the date of the Notice appealed against;
(c) the address of the premises concerned;
(d) the name and address of the respondent (i e the inspector);
(e) particulars of the requirements or directions appealed against; and
(f) the grounds for the appeal.

3.90 The appeal must be lodged within twenty-one days from the date of the service of the notice on the appellant, although the industrial tribunal may extend the time limit on application if it is satisfied that it was not reasonably practicable to bring the appeal earlier. The twenty-one day time limit runs from the date of the receipt of the Notice (*DH Tools Co v Myers*).

3.91 If the appeal is against the imposition of an Improvement Notice, the lodging of the appeal will automatically suspend the operation of the Notice until the appeal is disposed of. If the appeal is against the imposition of a Prohibition Notice which takes effect immediately or is deferred, the appellant may apply for it to be lifted pending the hearing of the appeal, but if he fails to do so, the Notice will take effect despite the fact that an appeal is pending. Since a Prohibition Notice which takes immediate effect can have serious repercussions on the employer's business, appeals against them are often heard as a matter of urgency, if necessary, the very next day (*Hoover Ltd v Mallon*).

3.92 Industrial tribunals have wide powers to deal with preliminary matters prior to the appeal. They can require further and better particulars of the application, grant discovery of documents, issue attendance orders compelling witnesses to attend and so on. As a rule, at least fourteen days notice of the date of the hearing is given, unless a speedier hearing can be arranged by agreement between the parties. The hearing will normally be in public, unless a party applies for it to be heard in private on grounds of national security or if there may be evidence the disclosure of which would be seriously prejudicial to the interests of the appellants undertaking other than its effect on collective bargaining. Either side may be represented by a solicitor, barrister, or by any other person whom he desires to represent him, including a trade union official or the representative of an employers' association. If written submissions are made, these must be sent to the industrial tribunal seven days prior to the hearing, and a copy must be sent at the same time to the other party.

3.93 Each side is entitled to make an opening statement, give evidence on oath, call witnesses, cross-examine witnesses from the other side, introduce documentary or other evidence, and make a closing submission. The tribunals do not appear (to date) to have found it necessary to appoint assessors to assist them, but they can, and do, visit the premises in order to make their own informed judgment (e g *Wilkinson v Fronks*).

3.94 The tribunal will then make its decision, which may be a unanimous one or by a majority. If the tribunal consists only of two members, the chairman has a casting vote. The decision may be given orally, or in writing after consideration, but it will always be promulgated in writing and a copy sent to each side.

Grounds for appeal

3.95 An appeal may be lodged on the ground that the inspector lacked the legal power to impose an Enforcement Notice. This may be because the premises are not covered by the relevant statutory provisions (*Dicker & Sons v Hilton*, above), or because the inspector has misinterpreted the statutory provision in question. But the fact that the employer has complied with the requirements previously laid down by some official authority is not by itself sufficient objection to a subsequent requirement made on grounds of health or safety. In *Hixon v Whitehead*, the appellant had received permission from the district Environmental Health Officer to store 4,000 kgs of liquified petroleum gas on the premises. As a result of complaints from local residents, another inspector issued a Notice limiting the holding to 500 kgs. The Notice was affirmed by the industrial tribunal. There was no question of estoppel arising against the local authority, for the tribunal had to consider the avoidance of serious injuries to employees and to other persons who may be affected. Similarly, in *Williamson Cliff Ltd v Tarlington* the company installed a tank containing 29,000 gallons of butane. Planning permission had been given for this after a fire hydrant had been installed. Subsequently, another inspector insisted that a spray system be installed. Although the industrial tribunal expressed sympathy with the company which had installed the correct system initially, only to find that they were now required to add another one even though there was no new knowledge in relation to safety, the Improvement Notice was confirmed. The overriding consideration was the safety of employees and other persons who were likely to be affected should an explosion occur.

APPEALS AGAINST IMPROVEMENT NOTICES

3.96 An Improvement Notice may be issued if there is a breach of a relevant statutory provision. This may be HSWA or other legislation or Regulation. The provision in question may be an absolute one, or prefaced by the requirement that something shall be done 'so far as is reasonably practicable'. In either case, the issues will be determined largely by the evidence which can be adduced.

3.97 Thus in *Murray v Gadbury* a farmer and his labourer used a rotary grass cutter for twenty-six years without incident. An inspector issued an Improvement Notice requiring the farmer to have the cutter guarded in accordance with the Agriculture Machinery Regulations 1962. It was argued that since the machine had been used for such a long time without accident, the notice was unnecessary. The industrial tribunal rejected the appeal and confirmed the Notice. The farmer appealed to the Divisional Court, claiming that the law only required him to do that which was reasonably practicable. Again, the appeal was dismissed. The provisions of the Regulations were quite clear and were mandatory. On the other hand, in *Associated Dairies v Hartley*, the appellants, who used roller trucks, provided safety footwear for employees at cost price. One day the wheel of a truck ran over the foot of an employee, causing a fracture. An inspector issued an Improvement Notice requiring the employers to provide suitable safety footwear free of charge. An appeal against this Notice was allowed. There was no statutory requirement that such footwear should be provided, and the obligation under section 2 of HSWA to make arrangements for securing safety at work is subject to the limitation 'so far as is reasonably practicable'. In determining whether or not a requirement was reasonably practicable, it was proper to take into account the time, trouble and expense of the requirement and to see if it was disproportionate to the risks involved. In this case, it would cost £20,000 in the first year to provide the boots, and £10,000 each year thereafter. On the other hand, the likelihood of such an accident occurring was fairly remote, and there was no evidence that employees would use the boots if they were provided free. The tribunal concluded that the present arrangement whereby the employers made safety footwear available was satisfactory.

3.98 The fact that employees are content with existing arrangements is also irrelevant. In *File Tile Distributors v Mitchell* the firm had a cold water supply, with a gas ring and kettle on which they could heat water. Employees used this method without complaint, but an Improvement Notice requiring the firm to provide running hot water was upheld. The statutory requirements are designed to improve facilities for employees, and their acquiescence in lower standards must be discounted.

3.99 Nor is it relevant that there has not been a previous accident or dangerous occurrence. In *Sutton & Co Ltd v Davies* the inspector issued an Improvement Notice requiring a machine to be fenced. The company produced evidence that there had not been an accident arising from this particular machine in twenty-seven years. Nonetheless, the notice was affirmed. The requirements of the statute were absolute. One need not wait for an accident to happen before condemning a system as being dangerous or unsafe.

3.100 The statutory requirements are not met by providing substitutes, unless some other equally efficacious method is permitted by the legislation. In *Belhaven Brewery v McLean* the inspector issued an Improvement Notice requiring the company to securely fence transmission machinery by the use of an interlocking device attached to the doors or gates. This would switch off the power when the doors were opened. The company argued that this would be very expensive, and they wanted to deal with the problem by erecting safety screens. They would also put up a notice warning employees of the danger. Since the employees were of sufficient intelligence to see that the gates were in position while the plant was working, and since there was a high level of supervision, they wanted the Notice cancelled or modified. The Notice was confirmed without modification. The requirements of section 13 of the Factories Act could not be met by erecting a screen. Moreover, under section 2(1) of HSWA it was reasonably practicable to fit the interlocking device. The sacrifice in terms of cost was not disproportionate to the risk and dangers.

3.101 Industrial tribunals are more willing to allow appeals to the extent that the appellant requires more time to comply with the requirements of the Notice. In *Campion v Hughes* an Improvement Notice was issued requiring the employer to make changes to the means of access in the event of fire. It was held that the appeal would be allowed only for the purpose of extending the time for compliance. One of the requirements was that the fire escape be made on land which belonged to the local authority, and consent for doing the work had not been given. An extension of time for a further three months was given in order to enable the employer to obtain the appropriate consent.

3.102 Similarly, in *Porthole Ltd v Brown* a firm enlarged its kitchen on the ground floor, and this took away the stairways which led to the first floor lavatory used by employees. An outside stairway was erected to provide the necessary access to the lavatory, and an inspector issued an Improvement Notice requiring this to be covered. The operation of the Notice was suspended to permit the firm to obtain the consent of the landlord and planning permission from the local authority.

3.103 An unusual extension of time arose in the case of *Cheston Woodware Ltd v Coppell*, where an Improvement Notice was issued requiring the appellant to fit an exhaust appliance to a planing machine used for thicknessing. The firm pointed out that if the machine was being used for thicknessing and surfacing, no such exhaust appliance would be required by the Woodworking Machine Regulations 1964. The tribunal agreed that this was somewhat odd, and postponed the operation of the Notice to enable the appellant to apply for an Exemption Certificate.

3.104 As a general rule, an appeal cannot succeed merely because the employer is unable financially to comply with the requirements. In *Harrison (Newcastle-under-Lyme) Ltd v Ramsey* an Improvement Notice was issued requiring the company to clean and paint its walls in accordance with section 1(3) of the Factories Act. The company appealed on the ground that it could not afford to spend the money on the work, in view of its grave financial position. The Notice was confirmed by the tribunal. To hold otherwise would enable an employer to ignore the statutory requirements because of expense and undercut his competitors who were so complying.

3.105 However, the industrial tribunals are not totally unsympathetic with the financial plight of firms, and will take into account the record of compliance in the past. In particular, they are more likely to postpone (as opposed to cancel) the operation of an Improvement Notice where the matter concerns a 'health' rather than a 'safety' aspect (*R A Dyson & Co Ltd v Bentley*).

3.106 An Improvement Notice may be successfully challenged on the ground that the employer is in fact complying with the statutory requirements. In *Davis & Sons v Leeds City Council* the tenancy of a flat above a small bakery shop was subject to a condition that employees at the shop could use the toilet facilities at all times. An Improvement Notice was issued requiring the shop occupiers to provide readily accessible sanitary facilities instead of this arrangement, in order to comply with section 9(1) of OSRPA. The industrial tribunal allowed an appeal against the Notice. Section 9(5) of the Act recognised that facilities might have to be shared with others, and in a circular addressed to local authorities the (then) Ministry of Labour had stated that the effect of section 9(5) would be that 'workers in a lock-up shop might have to use the conveniences and facilities in adjacent premises'. The tribunal, having visited the premises, concluded that the toilet facilities in the flat were conveniently accessible, and cancelled the Notice.

3.107 A similar result was reached in *Alfred Preedy & Sons Ltd v Owens*, where an Improvement Notice was served alleging that a stone stairway leading to a storage room was in a dangerous condition. The tribunal cancelled the Notice after hearing evidence from a witness with a long experience in property management and maintenance that the defects were minimal and of no practical significance.

3.108 A Notice may be successfully challenged on the ground that the inspector has misunderstood the application of the statutory provision. In *NAAFI v Portsmouth City Council* an Improvement Notice was served requiring the appellants to maintain a constant temperature of 55 degrees, so as to conform with section 6(1) of OSRPA. The appellants argued that as the premises were used for the storage of fresh food, a temperature of between 41 and 50 degrees was adequate. The tribunal noted that section 6(3)(b) of the Act provides for an exception where it is a room in which the maintenance of a reasonable temperature would cause the deterioration of goods, provided employees had conveniently accessible means of keeping themselves warm. Since a warm room was provided, together with suitable clothing, the Notice was cancelled.

3.109 Nor may an inspector impose a non-statutory requirement. In *Chethams v Westminster Council* a Notice was issued because the appellants were allegedly in breach of Regulation 7(1) of the OSRPA (Hoists and Lifts) Regulations. The Notice required that the latest British Standards for lifts be adopted. This requirement was struck out, because British Standards is merely a guide for new work, and is not a statutory provision.

3.110 The fact that the breach in question is a trivial one is irrelevant. In *South Surbiton Co-operative Society v Wilcox* an Improvement Notice required the employers to replace a wash basin which was cracked. On appeal, the Notice was confirmed. The surface of the basin was not 'impervious' as required by the Washing Facilities Regulations 1964, and consequently it was not 'properly maintained' in accordance with section 10(2) of OSRPA. That the infringement was trivial was irrelevant to the validity of the Notice.

3.111 Nor does the validity of the Notice depend on the instructions for remedying the defect being precise. In *Chrysler (UK) Ltd v McCarthy* two Improvement Notices were issued following a fire at the company's premises. Appeals were lodged on the ground that one of the Notices was imprecise. On a preliminary point of law, the tribunal dismissed the appeal, and this was confirmed by the Queen's Bench Divisional Court. It was pointed out that industrial tribunals have wide powers under section 24 of HSWA to modify the requirements of the Notice, as they think fit in the circumstances. However, when the matter was returned to the tribunal, it felt that they lacked sufficient information on which to make such a modification, and the Notice was suspended to permit the parties to agree between themselves what requirements should be laid down.

3.112 When modifying the Notice, an industrial tribunal may add to the inspector's requirements as well as vary them (*Tesco Stores Ltd v Edwards*), but there is no power which enables the industrial tribunal to amend a Notice so as to include further allegations that an employer is in breach of other provisions of the Act (*British Airways Board v Henderson*).

APPEALS AGAINST PROHIBITION NOTICES

3.113 A Prohibition Notice may be issued if the inspector considers that there is a risk of serious personal injury, irrespective of whether or not there is a breach of any relevant statutory provision. Consequently, a certain amount of subjectivity is involved in the formation of such opinion, and accordingly this assessment can be challenged. In *Nico Manufacturing Co Ltd v Hendry* a power press operated by the company was examined and tested by a competent person who found certain defects. In accordance with Power Presses Regulations 1965 he made a report to the company and to the inspector of factories. A few weeks later, a director of the company operated the press, and when this came to the attention of the inspector, he issued a Prohibition Notice. An appeal was lodged on the ground that the worn state of the press did not constitute a likely source of danger to the employees, and that the deprivation of the use of the machine would cause a serious loss of production and endanger the jobs of other employees. It was held that the Notice would be confirmed. The industrial tribunal preferred the evidence of the expert witness called on behalf of the inspector in so far as his evidence as to the danger was in conflict with that of the company's technical director. Although there was a small likelihood that the press would break up, there was a danger that parts would fracture which would constitute a serious danger to operatives.

3.114 However, if a machine or process has been in use for a long time without any history of accident or injury, it may be easier to challenge the inspector's view that there is an imminent risk of serious personal injury. In *Brewer & Sons v Dunston* an inspector issued a Prohibition Notice on a hand-operated guillotine. The company had used the machine for eighteen years without incident, and also had nine other similar machines. The tribunal, having visited the premises and seen the machine in operation, were satisfied that there was no imminent risk of serious personal injury, and cancelled the Notice.

3.115 The industrial tribunals take a similar attitude to appeals based on expense or a request for an extension of time as they do with Improvement Notices. Thus in *Otterburn Mill Ltd v Bulman* the company operated four machines with no guards. After making a number of visits to the premises, the factory inspector insisted that the appropriate guards be fitted, and when this was not done, he issued deferred Prohibition Notices requiring the guards to be fitted within three months. The company appealed against the time limit imposed, and requested that they should be allowed to fence one machine every six months, as they did not have the necessary finance to make the improvements. This argument was rejected by the tribunal. It would not be right to insist that a prosperous company should do the work in a short time, while a struggling company should be given a much longer period. However, since it was a deferred Notice, in that the risk was not imminent, the time taken to put the matter right was always a factor to be taken into account. As the last (and only) accident recorded at the factory was about nine years ago, the tribunal was prepared to grant an extension of time in respect of one machine, in order to avoid serious embarrassment to the company.

Use of Enforcement Notices

3.116 The number of Notices issued in recent years is as follows:

	Prohibition Notices	*Improvement Notices*	*Total*	*Crown Notices*
1978	3,434	12,217	15,651	34
1979	3,674	13,517	17,191	29
1980	3,237	12,517	15,863	35

Application for review

3.117 An application may be made within fourteen days after the promulgation of the decision to an industrial tribunal to review its decision, on the ground that:

(a) the decision was wrongly made as a result of an error on the part of the tribunal staff;
(b) a party did not receive notice of the proceedings;
(c) the decision was made in the absence of a party;
(d) new evidence has come to light since the making of the decision, the existence of which was not previously known or foreseen;
(e) the interests of justice require a review.

3.118 It must be remembered that an application for review is not an appeal against the decision of the industrial tribunal, and thus it must fall strictly within one of the above five grounds. The application may be refused by a Chairman of tribunals sitting on his own if he thinks that it stands no reasonable prospect of success. In making the application, the appellant should state the facts or evidence upon which he seeks to base his case. If the tribunal decides to hear the application, it may vary or revoke the original decision and order a rehearing, or dismiss the application.

Costs

3.119 Unlike other proceedings before industrial tribunals, costs are normally awarded against the loser to the party who wins the case. The amount awarded may be a specific sum, or, in default of agreement between the parties, the amount may be taxed in accordance with the County Court scales, as directed. However, the award of costs is always a matter for the discretion of the tribunal. Thus in *South Surbiton Co-operative Society v Wilcox* (see para 3.110) an Improvement Notice was confirmed on appeal, but because the breach in question was a trivial one, the industrial tribunal refused to make an order for costs.

Further appeals

3.120 An appeal from a decision of an industrial tribunal which relates to an Enforcement Notice can only be made to the Queen's Bench Divisional Court on a point of law.

Power to deal with imminent danger (section 25)

3.121 Where an inspector finds any article or substance in any premises which he has power to enter, and has reasonable cause to believe that, in the circumstances in which he finds it, the article or substance is a cause of imminent danger of serious personal injury, he may seize it and cause it to be rendered harmless (whether by destruction or otherwise). Before doing so, if it is reasonably practicable to do so, he must take a sample and give it to a responsible person at the premises where it was found, marked in a manner sufficient to identify it. After the article or substance has been seized and rendered harmless, the inspector will prepare and sign a written report giving particulars of the circumstances in which the article or substance was seized and dealt with by him, and shall

(a) give a signed copy to a responsible person at the premises where it was found, and
(b) unless that person is the owner, give a copy to the owner. If the inspector cannot ascertain the name or address of the owner, the copy will be given to the responsible person in question.

Prosecutions for criminal offences (sections 33–42)

3.122 Historically, the various branches of the inspectorate have always sought to ensure compliance with health and safety legislation by giving advice and using persuasion, rather than compulsion. Prosecutions for offences are generally used as the weapon of last resort, and a decision to prosecute will frequently be influenced by such factors as the previous record of the person, the number of visits made by an inspector to secure compliance, the gravity of the incident (if one has occurred) the public interest in prosecuting for the offence, and so on. In England and Wales, a prosecution in respect of an offence under any of the relevant statutory provisions may only be instituted by an inspector, or by or with the consent of the Director of Public Prosecutions (section 38), and an inspector, if authorised by the enforcing authority, may conduct the prosecution before a magistrates' court even though he is not a solicitor or a barrister (section 39). In Scotland prosecutions are undertaken by the Procurator Fiscal.

3.123 Prosecutions under HSWA may be brought in respect of the following offences:

Offence	*Summary conviction*	*On indictment*
1. Failure to discharge a duty under ss. 2–7	£1,000	Fine
2. Contravening ss. 8–9	£1,000	Fine
3. Contravening Health and Safety Regulations	£1,000	Fine
4. Contravening any requirement made by Regulations relating to investigations or enquiries made by the Commission, etc under s. 14, or obstructing anyone exercising his powers	£1,000	
5. Contravening any requirement under s. 20 (powers of inspectors)	£1,000	
6. Contravening any requirement under s. 25 (power of the inspector to seize and render harmless articles or substances likely to cause imminent danger)	£1,000	Fine
7. Preventing a person from appearing before an inspector or from answering questions under s. 20(2)(j) (examinations and investigations)	£1,000	

Offence	*Summary conviction*	*On indictment*
8. Contravening a requirement or prohibition imposed by an Improvement Notice	£1,000 £100 per day in respect of continuing offences, unless the court has given time to remedy under s. 42(3)	Fine
9. Contravening a requirement of prohibition imposed by a Prohibition Notice	£1,000 £100 per day in respect of continuing offences, unless the court has given time to remedy under s. 42(3)	Fine and/or 2 years imprisonment
10. Intentionally obstructing an inspector	£1,000	
11. Contravening a notice served by the Commission under s. 27(1) requiring information	£1,000	Fine
12. Using or disclosing information in contravention of s. 27(4) (disclosure by the Crown or certain Government agencies of information to the Commission or Executive)	£1,000	Fine and/or 2 years imprisonment
13. Disclosure of information obtained under s. 27(1) or pursuant to any statutory provision, not within the exceptions of s. 28	£1,000	Fine

Offence	*Summary conviction*	*On indictment*
14. Making a false or reckless statement in purported compliance with a statutory provision, or for the purpose of obtaining the issuance of a document under any statutory provision	£1,000	Fine
15. Intentionally making a false entry in any register, book, or other document required to be kept, or to making use of such entry, knowing it to be false	£1,000	Fine
16. Forging document, or, with intent to deceive, using a forged document	£1,000	Fine
17. Pretending to be an inspector	£1,000	
18. Failing to comply with an order of the court under s. 42 (order to remedy)	£1,000 £100 per day in respect of continuing offences, unless the court has given time to remedy under s. 42(3)	Fine
19. Acting without a licence which is necessary under a relevant statutory provision	£1,000	Fine and/or 2 years imprisonment
20. Contravening the terms of such licence	£1,000	Fine and/or 2 years imprisonment
21. Acquiring, using or possessing explosives contrary to the relevant statutory provisions	£1,000	Fine and/or 2 years imprisonment

BURDEN OF PROOF (SECTION 40)

3.124 As a general rule, it is for the prosecution to prove its case beyond reasonable doubt. However, section 40 provides that if the offence consists of a failure to comply with a duty or requirement to do something 'so far as is practicable' or 'so far as is reasonably practicable' or to 'use the best practicable means', it shall be for the accused to prove that it was not practicable or not reasonably practicable to do more than was in fact done or that there was no better practicable means than was in fact used to satisfy the duty or requirement. However, when this burden is placed on the accused, he need only satisfy the court on the balance of probabilities that what he has to prove has been done (*R v Carr-Briant*).

OFFENCES DUE TO THE FAULT OF ANOTHER PERSON (SECTION 36)

3.125 Where the commission of an offence by any person is due to the act or default of some other person, that other person shall be guilty of an offence, and may be charged and convicted whether or not proceedings are taken against the first-mentioned person (section 36(1)). If an offence is committed by the Crown, but the Crown cannot be prosecuted (see para 3.9) and the offence is due to the act or default of a person other than the Crown, that person shall be guilty of the offence and may be charged and convicted accordingly (section 36(2)). Thus employees of the Crown may be convicted, even though the Crown itself is immune.

OFFENCES BY DIRECTORS, MANAGERS, SECRETARIES, ETC (SECTION 37)

3.126 Where an offence committed by a body corporate is proved to have been committed with the consent or connivance of, or attributable to the neglect on the part of, any director, manager, secretary or other similar officer of the body corporate, or a person who was purporting to act in such capacity, then he, as well as the body corporate shall be guilty of that offence and shall be liable to be proceeded against (section 37(1)). This section was considered in *Armour v Skeen*, where the Strathclyde Regional Council and its Director of Roads were both prosecuted for a breach of safety Regulations, lack of a safe system of work, and failing to notify an inspector that certain work was being undertaken. As a result of these failures, an employee of the Council was killed. The alleged neglect on the part of the Director of Roads was a failure to have a sound safety policy for his department, failing to provide information to his subordinates, and a failure to provide training and instructions in safe working practices. He was convicted of the offences, and the conviction was upheld on appeal. The fact that section 2 of HSWA imposes a duty on the employers to provide a safe system of work did not mean that there was no duty on his part to carry out that duty. Section 37(1) refers to 'any neglect', not to the neglect of a duty imposed. The offences were committed by the body corporate, but were due to his neglect. Further, although his title as 'Director of Roads' did not mean he was a 'director' within the meaning of section 37(1), he was within the ambit of the words 'manager . . . or similar officer'.

3.127 Persons who purport to act as directors, managers, secretaries or similar officers are equally liable. Thus if a person acts as a director even though he has been disqualified from doing so under the Companies Act, he is purporting to act as such. Nor do the words 'purporting to act' imply that there is a fraudulent or false intention in so acting. Anyone who acts in a managerial capacity may be held liable under section 37(1) whatever title he may have. If the affairs of the body corporate are being managed by its members (e g a workers' co-operative), then the acts of a member which are in connection with his managerial functions are within the meaning of this section (section 37(2)).

ADDITIONAL POWERS OF THE COURT (SECTION 42)

3.128 If a person is convicted of an offence under any of the relevant statutory provisions in respect of any matter which appears to the court to be something which is in his power to remedy, the court may (in addition to or instead of any other punishment) order him, within such time as may be fixed, to take specified steps to remedy the matter. An application may be made to the court for an extension of time within which to comply with the order.

3.129 If a person is convicted of an offence under section 34(4)(c) (acquiring, possessing or using an explosive article or substance in contravention of a relevant statutory provision), the court may order the article or substance to be forfeited or destroyed or dealt with in such other manner as the court may order. Before making a forfeiture order the court must give an opportunity to the owner (or any other person with an interest in the article or substance) an opportunity to show cause why the order should not be made.

3.130 A failure to comply with an order under section 42 (e g if a person fails to take the necessary remedial action) is punishable by a fine of up to £100 per day. Additionally, it may amount to a contempt of court.

TIME LIMITS FOR PROSECUTIONS

3.131 By virtue of section 127 of the Magistrates' Courts Act 1980 a prosecution for a summary offence must be commenced by the laying of an information (i e issuing a summons) within six months from the date of the commission of the offence. However, section 34 of HSWA specifies that in certain cases an extension of the time limit may be possible. These are:

(a) where there has been a special report made by a person holding an investigation under section 14(2)(a) (see para 2.8);
(b) where a report is made by a person holding an enquiry under section 14(2)(b) (see para 2.9);
(c) a coroner's inquest is held touching the death of a person which may have been caused by an accident happening at work, or a disease contracted at work;
(d) a public enquiry is held into a death so caused under Scottish legislation;

and it appears from the report, inquest or inquiry that a relevant statutory provision was contravened, then summary proceedings may be commenced at any time within three months from the making of the report, or the conclusion of the inquest or inquiry (section 34(1)).

3.132 If an offence is committed by a designer, manufacturer, importer or supplier (i e section 6 offences) then summary proceedings may be commenced at any time within six months from the date when the enforcing authority had sufficient evidence, in its opinion, to justify a prosecution (or, in Scotland, to justify the making of a report to the Lord Advocate). A certificate of an enforcing authority stating that such evidence came to its knowledge on a specified date shall be conclusive evidence of that fact (section 34(3), (4), (5)).

3.133 When an offence is committed by reason of a failure to do something within a fixed time, the offence shall be deemed to continue until that thing be done, and time will not run until then (section 34(2)).

INDICTABLE OFFENCES

3.134 Because of the maxim 'Time does not run against the Crown' a prosecution in respect of an indictable offence is never barred by time limits. Since most of the offences under HSWA are triable either way (i e potentially indictable), the six months time limit has limited effect.

VENUE (SECTION 35)

3.135 An offence under any of the relevant statutory provisions may be treated as having been committed at the place where the plant or substance is for the time being, for the purpose of conferring jurisdiction on any court to entertain proceedings for an offence.

4 Health, safety and welfare in factories

4.1 It will be recalled (chapter 1) that the original purpose of factory legislation was to improve working conditions by means of the threat or application of criminal sanctions, although this receded into the background as civil claims increased which were based on a breach of a statutory duty. Some doubts have been expressed as to whether or not a civil claim will lie in respect of every provision of the Factories Act (particularly those which deal with welfare matters) but the point need not detain us here. Other legal problems which arise out of claims for injuries stemming from a breach of the Act and arising at common law will be dealt with in chapter 9.

Application of the Act

4.2 It will also be recalled that Regulations will be made under HSWA which will refer to activities, rather than to premises, and thus the provisions of the Factories Act will gradually be replaced. Until this process is completed, the Factories Act 1961 only applies to those premises which in law constitute factories. Hence, the duties are generally placed on the 'occupier', rather than upon the employer, although in practice this will usually be the same person. An occupier is someone who runs the factory and who regulates and controls the work that is done there (*Ramsay v Mackie*). Some of the duties in the Act are placed on the owners of the premises (e g where separate parts of a building are let off to different tenants) and he will be responsible for those matters which are under his control, for example, the common parts of stairways and passages.

4.3 The Act is designed to protect all persons who work in a factory, whether or not they are employed by the owner, occupier or employer, or are employed by some other person, or are self-employed. However, Regulations under the Act may expressly or by implication only apply to a restricted class of person (*Canadian Pacific Steamship Ltd v Bryers*).

4.4 A number of Regulations have been made under the Act (and its predecessors) which are still in force, and which may modify the Act to a material extent. In *Miller v William Boothman*, a worker was injured when using a circular saw which was fenced in accordance with the Woodworking Machinery Regulations 1922. He argued that he was entitled to rely on the provisions of section 14 of the Factories Act which are absolute in their requirement that dangerous parts of machinery must be securely fenced. It was held that he could not recover damages under section 14. The power was given to the Minister to modify the Act, and the Regulations prevailed. Similar modifications of statutory provisions which apply to particular types of machinery can be found in the Abrasive Wheels Regulations 1970 (passed as a result of the decision in *J Summers & Sons v Frost*) and the Power Press Regulations. However, the fact that Regulations exist do not absolve the occupier from taking proper steps in relation to those matters for which the Regulations make no provision (*Automatic Woodturning Co v Stringer*).

What is a factory?

4.5 Section 175(1) defines a factory as follows:

> any premises in which, or within the close or curtilage or precincts of which, persons are employed in manual labour in any process for or incidental to any of the following purposes, namely:—
> (a) the making of any article or part of any article, or
> (b) the altering, repairing, ornamenting finishing, cleaning or washing or the breaking up or demolition of any article, or
> (c) the adapting for sale of any article, or
> (d) the slaughtering of animals,
> (e) the confinement of such animals while awaiting slaughter (not being a cattle market),
>
> being premises in which the work is carried on by way of trade or for the purposes of gain and to or over which the employer of the persons employed therein has the right of access or control.

4.6 This somewhat extensive statutory definition requires further examination, for although it may be easy to describe a factory, it is not easy to define one, and in practice many marginal situations occur. It is tempting to say that common sense should be applied in any particular situation, but the literal wording of the Act needs to be considered first.

4.7 (1) There must be manual labour involved. This appears to mean 'working with one's hands' although the degree of strength is irrelevant. Nor is the degree of skill significant, as long as working with one's hands is the main or predominant activity. In *Joyce v Boots Cash Chemists* a porter carried parcels into a chemists shop. Although he was engaged in manual labour, the premises were not a factory, for his work was not 'for or incidental to' the processes of a factory. In *Hoare v Robert Green Ltd* a girl who made wreaths, crosses and bouquets in a room behind a florist's shop was held to be working in manual labour and consequently the premises were a factory, for she was engaged in making an article, and the quantum of labour involved was irrelevant. In a leading case in which the authorities were examined, it was suggested that there are a large number of people who work with their hands (authors, painters, archaeologists, art restorers etc), but who are really engaged in intellectual activities, and the manual labour aspect is largely incidental. In this case (*J & F Stone Lighting and Radio Ltd v Haygarth*) a radio and television engineer diagnosed and repaired faults in radio and television sets in a room behind a shop, and it was held that he was working in manual labour, and the premises were a factory.

4.8 (2) An 'article' is anything corporeal i e any commodity in bulk, whether solid, liquid or gaseous in form. Thus water is an article (*Longhurst v Guildford, Godalming and District Water Board*), as is coal gas (*Cox v Cutler & Sons Ltd*), but not a live animal (*Fatstock Marketing Corporation v Morgan*). Electricity stations are subject to their own special provisions in the Act (section 123).

4.9 (3) The phrase 'adapting for sale' is also a question of fact and degree. Thus packing sweets (*Fullers Ltd v Squire*) bottling beer (*Hoare v Truman, Hanbury Buxton & Co*) and cutting timber up (*Smith v Supreme Wood Pulp Co Ltd*) have all been held to be activities which were adapting an article for sale. But the mere testing of an article is not adapting for sale, even though it is done prior to the sale. Something must be done to the article in some way which makes it different from what it was before (*Grove v Lloyds British Testing Co Ltd*). Premises which are used for the pumping of water into peoples' homes are not a factory, as they are solely concerned with the distribution of the article, but a filtration plant is a factory, as it is adapting water for sale (*Longhurst v Guildford, Goldaming and District Water Board*).

4.10 (4) The processes must be carried on by way of trade or for the purposes of gain. This results in the exclusion of activities carried out in the workroom of a prison (*Pullen v Prison Commissioners*), instruction classes carried out in an educational institution (*Weston v London County Council*), the kitchen of a State run hospital, etc. However, activities carried on by the Crown and municipal authorities are within the Act even though they may not be carried on by way of trade or for purposes of gain, in the sense that they are not profit-making (section 175(9)).

4.11 (5) An open air site may constitute a factory, but there must be some geographical boundaries, even if these are not walls or fences (*Barry v Cleveland Bridge and Engineering Co Ltd*).

4.12 (6) The premises must be used for or incidental to the purposes of a factory. Thus, although a whole area may constitute a factory, it is possible that internal parts of premises within the curtilage of a factory are not part of the factory. In *Thomas v British Thomson-Houston Co Ltd*, within a factory there was a restaurant which was used by the directors, while the rest of the workforce used a works canteen. A worker was injured while cleaning the windows of the restaurant, and he brought a claim based on a failure to provide a safe means of access and a safe place at which to work (section 29, below). It was held that although the restaurant was within the curtilage of the factory, is was not used for, or incidental to, the purposes of the factory, and the claim based on a breach of section 29 failed. However, in *Luttman v ICI Ltd*, an industrial canteen, used by the workforce, was held to be premises which were incidental to the purposes of a factory, and was within the definition of section 175.

4.13 Section 175(2) goes on to specify a number of other premises which are to be regarded as factories, even though they do not come within the above definition. These include:

(a) any yard or dry dock in which ships are constructed, repaired, refitted, finished or broken up;
(b) any premises in which the business of sorting any articles is carried on as a preliminary to work carried on in a factory;
(c) any premises in which the business of washing or filling bottles or containers or packing articles is carried on incidentally to the purposes of the factory;

(d) any premises in which the business of hooking, plaiting, lapping, making up or packing of yarn or cloth is carried on;
(e) any laundry carried on as an ancillary to another business or incidentally to the purposes of any public institution;
(f) any premises in which the construction, reconstruction, or repair of locomotives, vehicles or other plant for use for transport purposes is carried on as ancillary to a transport undertaking or other industrial or commercial undertaking;
(g) any premises in which printing by letterpress, lithography, photogravure, or similar process, or bookbinding is carried on by way of trade or for purposes of gain or incidentally to another business carried on;
(h) any premises in which the making, adaptation, or repair of dresses, scenery or properties is carried on incidentally to the production, exhibition or presentation by way of trade or for purposes of gain of cinematograph films or theatrical performances (not being a stage or dressing room of a theatre in which only occasional adaptations or repairs are made);
(i) any premises in which the business of making or mending nets is carried on incidentally to the fishing industry;
(k) any premises in which mechanical power is used in connection with the making or repair of articles of metal or wood incidentally to any business carried on by way of trade or for purposes of gain;
(l) any premises in which the production of cinematograph films is carried on by way of trade or for purposes of gain;
(m) any premises in which articles are being made or prepared incidentally to the carrying on of building operations or works of engineering construction, not being premises in which such operations or works are being carried on;
(n) any premises used for the storage of gas in a gasholder having a storage capacity of not less than 5,000 cubic feet.

Cleanliness (section 1)

4.14 Every factory shall be kept in a clean state, free from effluvia arising from any drain, sanitary convenience or nuisance. In particular, accummulations of dirt and refuse shall be removed daily by a suitable method from the floors and benches of workrooms, and from the stairs and passages, and the floor shall be cleaned at least once a week by washing or, if it is effective and suitable, by sweeping or other method. All inside walls and partitions, and ceilings which have a smooth impervious surface shall be washed at least once in every fourteen months with soap and hot water or other method approved by the inspector, or, if they are painted or varnished, shall be repainted or revarnished at least every seven years and washed or cleaned every fourteen months or, in every other case, be whitewashed or colourwashed at least every fourteen months. The only exception arises when the factory does not use mechanical power and less than ten persons are employed there, unless the inspector requires otherwise. The Factories (Cleanliness of Walls and Ceilings) Order 1960 (as amended) lays down the standards for the repainting, revarnishing, whitewashing or colourwashing, and provides for exemptions in respect of certain types of premises.

4.15 Financial hardship cannot be pleaded successfully as an excuse for not conforming to the requirements of section 1. In *Harrison (Newcastle-under Lyme) Ltd v Ramsey* an Improvement Notice was issued requiring the company to clean or paint its walls in accordance with section 1(3). The company had been in occupation of the premises for three years, and appealed against the Notice on the ground that it could not afford to do the work, in view of its critical financial position. The industrial tribunal upheld the Notice. The financial embarrassment of the factory occupier was irrelevant. To take this into account would mean that an employer could deliberately ignore health and safety provisions on the ground of expense and thus undercut his competitors who were complying with the statutory requirements.

Overcrowding (section 2)

4.16 A factory shall not be so overcrowded as to cause the risk of injury to persons employed in it. The maximum number to be employed shall be such that each person has 400 cubic feet (11 cubic metres) (unless an exemption applies) and space above 14 feet from the floor is not to be taken into account for determining the cubic space available. A notice specifying the number of persons who may be employed in a workroom shall be displayed unless the inspector allows otherwise.

Temperature (section 3)

4.17 A reasonable temperature shall be provided and maintained. If a substantial proportion of the work is done sitting down and does not involve serious physical effort a temperature of less than 60 degrees F (16 degrees C) after the first hour shall not be deemed to be reasonable while the work is continuing. At least one thermometer shall be provided and maintained in a suitable position. A failure to provide an adequate temperature was held to be a breach of an implied term of the contract of employment in *Graham Oxley Tool Steels Ltd v Firth*, entitling the employee to resign and claim that she had been constructively dismissed (see chapter 10).

Ventilation (section 4)

4.18 Effective and suitable provision shall be made for securing and maintaining by the circulation of fresh air the adequate ventilation of the workroom, and for rendering harmless, so far as practicable, all such fumes, dust and other impurities generated in the process of the work carried on as may be injurious to health. This section should be read in conjunction with section 63, which deals with the removal of dust and fumes, and section 30, which deals with working in confined spaces where fumes are present.

4.19 In *Nicholson v Atlas Steel Foundry and Engineering Co Ltd* a factory process gave rise to a large amount of siliceous particles in the atmosphere. The only ventilation was provided by two doors. The employee contracted pneumoconiosis from which he died. It was held that the failure to ventilate the workroom must have exacerbated the hazard to which he was exposed, and the dust in his lungs must have materially contributed towards his death. Consequently the employers were held liable. It should be noted that section 4 is only concerned with the circulation of fresh air, and does not, for example, require that masks or respirators should be provided (*Ebbs v James Whitson & Co Ltd*).

4.20 The dust or fumes must have been caused by the work which is being carried on. If there is some other cause, the section will not apply. For example, in *Brophy v Bradfield & Co Ltd* an employee had gone into a boiler room (into which he had no right to go), closed the door, and opened the furnace for warmth. He was overcome by carbon monoxide fumes, and died. It was held that (a) the boiler room was not a workroom, and (b) the fumes were not generated in the course of any process or work carried on in the factory, and the employers were thus not liable for a breach of statutory duty (for the claim based on common law negligence, see chapter 9).

Lighting (section 5)

4.21 Effective provision shall be made for securing and maintaining sufficient and suitable lighting, whether natural or artificial, in every part of the factory in which persons are working or passing.

The Factory (Standards of Lighting) Regulations 1941 did lay down specific standards to be achieved, but these have been repealed, as it was thought that they were unnecessary. In *Thornton v Fisher and Ludlow Ltd* a cleaner employed by the defendants was walking along a factory road early in the morning. She tripped over a coil of wire and was injured. There were lights along the walls by the road way, but these were not switched on. It was held that the defendants were liable for her injury, for by not ensuring that the lights were turned on, they had failed to make effective provision for securing and maintaining sufficient lighting. However, if adequate lighting is provided, the fact that a shadow is cast over a piece of the work in progress thus reducing the amount of light within the area covered by the shadow does not constitute a breach of section 5 (*Lane v Gloucester Engineering Co Ltd*).

4.22 All glazed windows and skylights used for lighting purposes must be cleaned on the inner and outer surfaces and free from obstruction, although these may be whitewashed or shaded in order to mitigate against glare or heat.

Drainage of floors (section 6)

4.23 If a process is carried on which renders the floor liable to be wet, and this is capable of being removed by drainage, effective means for doing this must be provided and maintained.

Sanitary conveniences (section 7)

4.24 Sufficient and suitable sanitary conveniences shall be provided, maintained and kept clean, with effective lighting. Where both sexes are employed, there must be separate accommodation for each sex. The Sanitary Accommodation Regulations 1938 lay down certain minimum standards to be observed. There must be one convenience for every twenty-five females, and one for each twenty-five males (not merely suitable as a urinal), although if there are more than 100 males and there are sufficient urinals, it is only necessary to have four conveniences for the first 100 males and one for each forty thereafter. If there are more than 500 males, one convenience for every sixty males is sufficient.

4.25 It will be noted that the Act requires separate toilet facilities for men and women. The fact that there are no such separate facilities is not an excuse for refusing to employ a man or woman, as the case may be. The Sex Discrimination Act 1975, section 7 makes it permissible to discriminate against a person on grounds of sex where the sex of the person is a genuine occupational qualification for the job, and one of these arises where the employee is required to live in residential accommodation provided by the employer, and there is no separate sleeping and sanitary accommodation, and it is not reasonable to expect the employer to equip those premises with such facilities, or to provide other premises for the different sexes (section 7(2) (c)). If an employer wishes to rely on this genuine occupational qualification he must show that his needs are truly genuine. In *Hermolle v Government Communications HQ* a woman applied for a job on the Ascension Islands but was refused the post because, it was alleged, there was no separate sleeping accommodation or sanitary facilities. However, the cost of providing these was fairly modest, and the employers failed to discharge the burden of showing that it was unreasonable of them to make such separate facilities available. Thus, it is unlikely that such a defence would excuse an alleged discriminatory act by a factory occupier based on the absence of separate toilet facilities.

4.26 The conveniences shall be properly screened, conveniently accessible and the interior shall not be visible to persons of the opposite sex when the door is opened. The separate conveniences for each sex shall be indicated by a suitable notice.

Medical examinations (section 10A)

4.27 If an Employment Medical Adviser (EMA, see chapter 2) is of the opinion that a person's health has been or is being injured (or that it is possible that he is or will be injured) by reason of the work he is doing, the EMA may serve a notice on the factory occupier requiring the occupier to permit a medical examination of that person to take place. The notice will state the time, date and place of the examination, which must be at reasonable times during working hours. Every person to whom it relates shall be informed of its contents and of the fact that he is free to attend for that purpose. If the examination is to take place in the factory, suitable accommodation shall be provided.

Medical supervision (section 11)

4.28 If it appears to HSE that in a particular factory cases of illness have occurred due to the nature of the work or substances used, or young persons are employed in work which may cause risk of injury to their health, or there is a risk of injury to the health of other persons employed from any substance or materials or other conditions, HSE may, by Order, require reasonable arrangements to be made for medical supervision of those persons. To date, the powers contained in this section do not appear to have been utilised.

General safety provisions (sections 12–16)

4.29 These sections have received more consideration in civil and criminal proceedings than any other sections of the Act. Clearly, this reflects the large number of accidents and potential injuries which are likely to result from the use of industrial machinery, and a determination by the legislature to minimise those risks and to achieve the highest possible safety standards. The five sections form a single Code, and may be interpreted as such (*Callow (Engineers) Ltd v Johnson*).

Prime movers (section 12)

4.30 Every flywheel connected to a prime mover, and every moving part of a prime mover shall be securely fenced. A prime mover is any appliance which provides mechanical energy derived from fuel, steam, water or any other source (section 176(1)). Electric generators, motors and rotary converters and flywheels directly connected thereto shall be securely fenced unless they are in such a position or of such construction as to be as safe to every person employed or working on the premises as they would be if securely fenced.

Transmission machinery (section 13)

4.31 Every part of transmission machinery shall be securely fenced unless it is in such a position or of such construction as to be as safe to every person employed or working on the premises as it would be if securely fenced. Transmission machinery means every shaft, pulley, wheel, drum, coupling, clutch, driving belt or other device by which the motion of the prime mover is transmitted to or received by any engine or appliance. This includes machinery not driven by mechanical power (*Richard Thomas & Baldwins Ltd v Cummings*), but there must be some thing which moves or revolves. Thus a hydraulic accumulator is not transmission machinery (*Weir v Andrew Barclay & Co Ltd*). An efficient device shall be provided and maintained in every room or place where work is being carried on, by which the power can be cut off from the transmission machinery in that room or place.

Other machinery (section 14)

4.32 Every dangerous part of any machinery (other than prime movers and transmission machinery) shall be securely fenced unless it is in such a position or of such construction as to be as safe to every person working or employed in the premises as it would be if securely fenced. In so far as the dangerous part cannot be securely fenced by a fixed guard, the section will be complied with if there is an automatic device which prevents the operator from coming into contact with that dangerous part (section 14(2)). Stock bars which project beyond the head-stock of a lathe shall similarly be securely fenced unless their position makes them as safe to every person working or employed as they would be if they were securely fenced.

4.33 The following points arise out of the interpretation of these sections.

i *What is machinery?*
4.34 The Act does not define this word, and a certain common-sense approach must be taken. 'Machinery' is a wider term than 'machine'. There is no requirement that the machinery must be driven by mechanical or electrical power, and it may include a large structure (*Quintas v National Smelting Co* – a cable-way) or a small hand-tool (*Close v Steel Co of Wales* – an electric drill). Nor is machinery excluded from the Act because it is mobile, for the Act is designed to protect people from the dangerous parts, not from the movement of machinery. In *British Railways Board v Liptrot*, a mobile crane was mounted on a four-wheeled chassis with rubber wheels. The respondent was injured when he was caught between the revolving body of the crane and the wheels. It was held that the crane was 'machinery' within the meaning of the Act. Mobility did not grant immunity, and the fact that cranes and similar equipment are subject to detailed provisions in section 27 did not exclude the operation of section 14. If vehicles contain dangerous parts, these must be securely fenced.

ii *Which machinery?*
4.35 The machinery must be part of the equipment of the factory. Thus, if lorries, car, etc come to visit the premises, they are not part

of the factory equipment, and hence not within the scope of the Act. Equally, machinery which is being made in the factory is not covered (*Parvin v Morton Machine Co Ltd*). But once machinery is installed in a factory, it is part of the equipment, even though it is only being tested for use (*Irwin v White Tomkins and Courage*). In *Thorogood v Van Den Burghs and Jurgens Ltd* an electric wall fan was taken from its position to an engineering workshop for repair. Since the latter premises were still part of the factory, it was held that the duty to fence the fan still applied.

iii *What is a 'dangerous part'?*

4.36 Some limit must be placed on this phrase, and it cannot be applied to every object which causes injury, otherwise the most harmless physical things would have to be fenced (chairs, tables, etc) which could hardly be described as being 'dangerous'. The modern test has been restated in *Close v Steel Co of Wales*. A part of machinery is dangerous if it is 'a reasonably foreseeable cause of injury to anyone acting in a way in which human beings may be reasonably be expected to act, in circumstances which may reasonably be expected to occur'. In other words, a degree of foreseeability is required, and this is largely (if not totally) a question of fact and degree. In *Mitchel v North British Rubber Co*, Lord Cooper gave a further explanation which has subsequently been cited with approval by the House of Lords (*John Summers v Frost*). 'The question is not whether the occupiers of the factory knew that it was dangerous; nor whether the factory inspector had so reported; nor whether previous accidents had occurred; nor whether the victims of these accidents had, or had not, been contributorily negligent. The test is objective and impersonal. Is the part, in its character, and so circumstanced in its position, exposure, method of operation and the like, that in the ordinary course of human affairs danger may reasonably be anticipated from its use unfenced, not only to the prudent, alert and skilled operative intent upon his task, but also to the careless and inattentive worker whose inadvertant or indolent conduct may expose him to the risk of injury or death from the unguarded part.' Thus, the fact that there was never been a previous accident, or that the factory inspector has never prosecuted, may have some relevance in evidence, but cannot be conclusive.

4.37 It is reasonable to foresee that workers will disobey instructions not to put their hands into an unfenced part, and that they will be careless in the way they operate the machine. In *Smith v*

Chesterfield and District Co-operative Society the plaintiff worked on a machine the rollers of which were protected by a guard which came down to within 3 inches of the bed of the machine. Despite instructions to the contrary, she placed her fingers under the guard and was injured. It was held that there was a breach of section 14(1), for her conduct, though careless, was foreseeable. It is also possible to foresee that a worker will be indolent, disobedient, or simply tired. In *Woodley v Meason Freer* employees were specifically warned not to put their hands into a machine for the purpose of removing obstructions. When a worker acted in contravention to this instruction, and was injured, it was held that the employers were guilty of an offence under the Act.

4.38 It is even possible to foresee that a workman will be stupid in his behaviour. In *Uddin v Associated Portland Cement Ltd* the plaintiff climbed a ladder in order to chase a pigeon which had flown into the roof of the factory. He slipped, and caught his clothing on an unfenced part of machinery and was injured. The employers were held liable. The Act is designed to protect persons who are 'employed or working' on the premises, and the fact that the plaintiff had been guilty of an act of folly was only relevant in assessing the extent of his contributory negligence.

4.39 On the other hand, there comes a point where no amount of foreseeability can prevent an accident. In *Rushton v Turner Asbestos Ltd* the employee was specifically told that he must not attempt to put his hand into a machine. Nonetheless he did so and was injured. Although this failure to securely fence may have constituted a criminal offence which would have warranted prosecution, in the civil proceedings which ensued it was held that he was the sole author of his misfortune, and the employers were not liable for breaking their statutory duty or for negligence. And in *Carr v Mercantile Produce Ltd* a girl forced her hand into a hole in a machine which was 3 inches in diameter. This act, of sheer perversity, was not foreseeable.

4.40 It is only possible to guard a machine against dangers which might reasonably be foreseen. It follows that if the situation is such that it can only be foreseen with the benefit of hindsight, it is not reasonably foreseeable. In *Burns v Joseph Terry Ltd* the plaintiff climbed some ladders in order to get some cocoa beans from a shelf. Just below the shelf were some cog wheels, which were part of the transmission machinery, and these were covered in front by a mesh guard. The ladder slipped, and somehow the boy got his hand behind the guard and was injured. It was held that the incident was not reasonably foreseeable. The transmission machinery was guarded against such dangers as were foreseeable.

4.41 Foreseeability is relevant to the conduct of the worker, not to the operation of the machine. Thus if a worker is injured on an unfenced part of machinery, and the injury would not have occurred if the part had been securely fenced, the employer will be liable even though the accident happened in an entirely unforeseen manner, e g if the machine makes an uncovenanted stroke (*Millard v Serck Tubes Ltd*).

iv *Danger in juxtaposition*

4.42 In several cases the problem has been posed as to whether or not a part of machinery is dangerous if the danger only arises because of a juxtaposition with the machine and materials used by the machine. The Act, after all, does not require the fencing of dangerous materials in the machine (*Bullock v John Power (Agencies) Ltd*). In *Eaves v Morris Motors* the plaintiff's hand was injured when it was caught on a sharp bolt being made on a moving machine tool, and it was held that the employers were not liable. However, in *Midland and Low Moor Iron and Steel Co Ltd v Cross* the House of Lords held that whether a part of machinery was dangerous must be determined by considering what the machine was designed to do under normal circumstances. In this case, there was a power-driven machine for straightening metal bars by squeezing them between rollers. An employee fed the bars into the rollers, which were not fenced. His attention was distracted, and his hand was nipped between the metal bar and the roller. The rollers themselves were not dangerous, as there was a sufficient gap between the moving lower and the stationary upper roller to escape any injury. The danger only arose because the bar was being inserted. It was held that the employers were in breach of the Act, for the danger arose from the normal operation of the machine. Nor does a part of machinery cease to be dangerous merely because it is stationary, if the material being used is moving, and thus causes a potential danger (*Callow Engineers Ltd v Johnson*). In other words, if there is a juxtaposition of parts of a machine and a workpiece being used by the machine which causes a danger, it must be fenced. In *Wearing v Pirelli Ltd* the plaintiff was injured when his hand came into contact with a harmless rubber fabric which was attached to a dangerous revolving drum which was not fenced. The employers were held liable for the injury. But a tool being used on a machine does not constitute a dangerous part of machinery, and there is no requirement to fence a danger which arises from the use of the tool (*Sarwar v Simmons and Hawker Ltd*).

4.43 There is no duty to fence against a danger arising from the juxtaposition of one piece of machinery and another, or a danger caused by the proximity of a moving part of machinery and some other extraneous object. In *Pearce v Stanley Bridges Ltd*, the plaintiff injured his arm when it was caught between the rising platform of a lifting machine and a conveyor belt. It was held that the Act did not impose an obligation to fence a gap between two machines. Moreover, the alleged danger was not reasonably foreseeable, and for both these reasons the claim failed.

v *When is machinery 'securely fenced'?*

4.44 A fence must not be a mere barrier (*Quintas v National Smelting Co*), but sufficient to protect the worker adequately. On the other hand, it does not cease to be secure because its protection can be circumvented and rendered useless by some act of perverted and deliberate ingenuity (*Carr v Mercantile Produce*, above).

4.45 Since the obligation is to fence, a substitute for fencing clearly constitutes a breach of the Act. In *Chasteney v Nairn* the employers put up a notice which read 'Do not put your hands in the machinery while it is in motion. Persons disregard this notice at their own risk.' The court had no doubt that the machinery was not securely fenced.

vi *The purpose of fencing*

4.46 A fence must prevent the worker from coming into contact with the dangerous part of machinery. It is not intended that a fence should protect him from parts of the machine which may fly out. In *Close v Steel Co of Wales*, the bit of a portable electric drill shattered and injured the plaintiff in his eye. Such shatterings were fairly common, but never before had a serious incident occurred. It was held that the employers were not liable for damages. Nor need the fence protect against flying parts of material being worked on the machine (*Nicholls v Austin (Leyton) Ltd*), though if an employer knows that parts of a machine or materials used by the machine are likely to fly out of the machine, this may give rise to liability for common law negligence, in that there may be a failure to ensure a safe system of working (see chapter 9) and an offence may have been committed under section 2 of HSWA.

4.47 If a worker is injured because a tool he is using is caught on a dangerous part of machinery, the employer will not be liable (*Sparrow v Fairey Aviation Ltd*), but in *Lovelidge v Anselm Odling Ltd* the plaintiff's tie was caught in a machine, as a result of which he was injured. It was held that the purpose of fencing was to guard against such contingencies, the clothing a worker wears is part of him for this purpose and employers were held liable.

vii *Protection for whom?*

4.48 In general, sections 12–16 are for the benefit of persons who are employed or working on the premises. This can mean employees, self-employed subcontractors, and so on. The fact that the employee is acting outside the scope of his employment is not relevant (*Westwood v Post Office*). For example, it was no part of the work of the plaintiff in *Uddin's* case (above) to chase pigeons, but nonetheless he was able to recover damages for a breach of section 14(1). However, if the employee is neither employed nor working in the legal sense, but 'doing a foreigner' in his spare time (even with the permission of the employer) the Act does not apply (*Napieralski v Curtis (Contractors) Ltd*).

4.49 It will be noted that section 14(2) (above) which permits the use of an automatic device instead of a secure fence is only expressed to be for the benefit of the operator.

4.50 The obligations in section 14 are absolute, and the fact that it is commercially impracticable or mechanically impossible to securely fence the machine (*Davies v Thomas Owen Ltd*), or to use the machine when securely fenced (*John Summers & Sons Ltd v Frost*) is immaterial. Equally, it is relevant that the dangerous part was fenced by the best known methods (*Dennistoun v Charles Greenhill Ltd*), or that the machine has been used in that manner for thirty years without any accident or complaint from the factory inspector (*Sutherland v James Mills Ltd*), although such factors may be of evidential value as to whether or not the parts of machinery were dangerous, or whether they were securely fenced (*Carr v Mercantile Produce*).

4.51 The fact that the injured person does not know or cannot explain adequately how the accident occurred is irrelevant once a breach of the statute has been shown. In *Allen v Aeroplane and Motor Aluminium Castings Ltd* the plaintiff was injured on a machine which was not fenced. The judge did not believe his account of the incident, and gave judgment for the employers, but this was reversed on appeal. Once a breach has been shown, there is a presumption that the accident would not have happened but for the breach.

4.52 The obligations in this part of the Act are placed on the factory occupier. In *Biddle v Truvox Engineering Ltd* an employee was injured as the result of a failure to securely fence a machine, and he sued his employers. They, in turn, joined, as a third party to the action, the manufacturer from whom they bought the machine, but the claim against them was dismissed. The decision is not without its critics, but nowadays the manufacturer or seller would be criminally liable under section 6(8) of HSWA, and the injured person would have his remedy under the Employer's Liability (Defective Equipment) Act or at common law under the doctrine of *Donoghue v Stevenson* (see para 9.3).

Unfenced machinery (section 15)

4.53 If machinery is in the process of examination, lubrication or adjustment, it may well be that it is not in a safe position or of safe construction. In these circumstances, the provisions of sections 13–14 will not apply if the examination, lubrication or adjustment can only be carried out while the part of machinery is in motion. In the case of transmission machinery being used in any specified process where, owing to its continuous nature, the stopping of it would seriously interfere with that process, the lubrication or mounting or shipping of belts shall be carried out in such methods and in such circumstances as may be prescribed.

4.54 The Operations at Unfenced Machinery Regulations 1938 (as amended) lay down the conditions which must be complied with when examining, lubricating or adjusting unfenced machinery whilst in motion where dangerous parts are exposed. The factory occupier must appoint a person to carry out this operation, who must be over the age of eighteen, sufficiently trained for the purpose, and aware of the dangers involved. He must be given a copy of the precautionary leaflet issued by HSC and, except for the setting of a machine by a toolsetter, he must be provided with a single piece overall. Another person instructed in the steps to be taken in an emergency must be within sight or hearing distance, and steps must be taken to ensure that other persons are not in a position as to be exposed to any risk of injury. These provisions do not apply if the machinery is being moved by hand or by an inching button.

4.55 The Regulations also deal with the lubrication, mounting, or shipping of belts in certain specified processes where this work cannot be deferred until the machinery is stopped (see the Schedule to the Regulations).

4.56 It will be noted that section 15 only provides a partial exemption from the stringent requirements of sections 13–14 when the work being carried on is that of examination, lubrication and adjustment, and even then only in the prescribed circumstances. Section 15 does not give any exemption in the case of the cleaning of a machine while it is in motion.

Construction and maintenance of fencing (section 16)

4.57 All fencing or other safeguards shall be of substantial construction and constantly maintained, and kept in position while the parts required to be fenced or safeguarded are in motion or use, except when they are necessarily exposed for examination and for any adjustment or lubrication shown by such examination to be immediately necessary. The phrase 'in use' means 'running as it was meant to run and doing the work it was meant to do' (*Richard Thomas and Baldwins Ltd v Cummings*). Thus if machinery is being repaired or being cleaned, it is not in use at the time. In *Knight v Leamington Spa Courier Ltd* a printing machine was being rotated slowly by an inching button. The object was to clean the machine during a slack period, and it was held that it was not 'in use' at the time. Acts which are preparatory to the machine being worked also do not make the machine in use (*Horne v Lec Refrigeration Ltd*) but in *Joy v News of the World*, paper was being threaded through the rollers of a printing machine which was running at its lowest speed, and it was held that the machine was in use at the time.

4.58 Even more difficult to understand is the phrase 'in motion'. Oddly enough, this does not mean 'in movement'! It means 'substantial movement of its normal workings, or . . . some movement reasonably comparable to its normal workings' (*Knight v Leamington Spa Courier Ltd*). Thus the slow, sporadic rotation or intermittent movement of machinery for the purpose of placing it in a more advantageous position in order to clean or repair it is not normal motion, whether this result is achieved by manpower or mechanical power. The distinction between motion and movement is one of fact and degree, to be determine by the judge. It has also been held that regard had to be made to the character of the movement as well as to its purpose (*Mitchell v W S Westin Ltd*). Thus a machine which is revolving rapidly must be fenced whether the purpose of the movement is production, examination, demonstration, adjustment, repairing, testing or whatever. In *Stanbrook v Waterlow & Sons Ltd* the cylinder of a printing machine was made to revolve at a high speed for a fraction of a second, and it was held that the machine was in motion. There is clearly a distinction between being moved slowly, and being in motion at a fast pace. If a machine is put into motion for its normal work, whether slowly or rapidly, it must be fenced. If the movement is slow, purposeful and deliberate, for the purpose of examination, lubrication or adjustment, the fencing provisions do not apply, but if this work is done when the machine is put into motion at a rapid pace, the sections will apply.

4.59 If an operation can be performed with the guards being kept in position, then it cannot be said that the dangerous parts of machinery were 'necessarily exposed' for any purpose, and the exemptions of section 16 will not apply (*Nash v High Duty Alloys Ltd*).

Construction and sale of machinery (section 17)

4.60 Every set-screw, bolt or key on any revolving shaft, spindle, wheel or pinion shall be so sunk, encased or otherwise effectively guarded as to prevent danger, and all spur or other toothed friction gearing which does not require frequent adjustment while in motion shall be completely encased, unless it is so situated as to be as safe as it would be if completely encased.

4.61 Any person who sells or lets on hire (or acts as an agent of a seller or hirer) who causes or procures the sale or hire of any machine for use in a factory which is intended to be driven by mechanical power which does not comply with this section shall be guilty of an offence. It will be recalled that in *Biddle v Truvox Engineering Co Ltd* the vendor of a machine was held not to be liable by virtue of this section to an injured workman as a joint tortfeaser along with the employer. However, if a machine has a defect which was negligently caused in the manufacture, an action for negligence would lie against the manufacturer at the instance of an injured person (*Hill v James Crowe Ltd*), and the actual employer of the injured person would be liable under the provisions of the Employers Liability (Defective Equipment) Act 1969 (see chapter 6). Additionally, the actual manufacturer etc would be liable under section 6 of HSWA (see chapter 3).

Dangerous substances (section 18)

4.62 Every fixed vessel, structure, sump or pit which contains any scalding, corrosive or poisonous liquid, the edge of which is less than 3 feet (920 millimetres) from the ground or platform from which a person might fall into it shall be securely covered or securely fenced to at least 3 feet (920 millimetres) above that ground or platform, or, if it is not possible to do this by reason of the nature of the work, all practicable steps shall be taken to prevent any person from falling in. This protection is against the nature of the liquid, which must be dangerous when a person comes into contact with it; it is not protection against drowning per se. Further, no ladder, stair or gangway shall be placed above, across or inside the vessel, etc, which is not at least 18 inches (460 millimetres) wide, securely fenced on both sides to a height of at least 3 feet (920 millimetres), and securely fixed.

4.63 There are special Regulations relating to fixed vessels in chemical works (Chemical Works Regulations 1922) and in relation to the use of vats in dyeworks (Kiers Regulations 1938).

Self acting machines (section 19)

4.64 A 'self acting' machine has a fixed part and a traversing part which moves backwards and forwards, and which may thus cause danger to someone trapped between these parts. Section 19 provides that if the traversing part (or any material carried on it) runs over a space over which any person is likely to pass, the traversing part, or materials on it, must not be allowed to go within 18 inches (460 millimetres) of any fixed structure which is not part of the machine. If the person in charge of the machine bone fide believes that another person is clear of the machine, he does not allow that other to pass a traversing part (*Crabtree v Fern Spinning Co Ltd*). (See further, Spinning by Self-acting Mules Regulations 1905.)

Cleaning of machines (section 20)

4.65 A woman or young person (i e below the age of eighteen) shall not clean any part of a prime mover or any transmission machinery while either is in motion, and shall not clean any part of any machine if the cleaning thereof would expose the woman or young person to the risk of injury from any moving part of that machine or any adjacent machinery. In *Taylor v Mark Dawson Ltd* a child was picking fluff from the rollers of a spinning machine which was in motion. The fluff had a resale value, and was not therefore waste, and the purpose of his activities was to prevent the rollers from becoming clogged. It was held that he was engaged in cleaning the machine. Further, the meaning of 'moving part' is not the same as 'in motion' (see section 16 above). In *Denyer v Charles Skipper and East Ltd* a seventeen-year-old boy was allowed to clean rollers on a printing machine which involved rotating them with an inching button. It was held that there was a breach of section 20.

Training and supervision of young persons (section 21)

4.66 No young person shall work at any prescribed machine unless he has been fully instructed as to the dangers arising and the precautions to be observed, and

(a) has received sufficient training in work at the machine, and
(b) is under adequate supervision by a person who has a thorough knowledge and experience of the machine.

4.67 The Dangerous Machines (Training of Young Persons) Order 1954 specifies the dangerous machines in question. These are:

(a) *machines worked with the aid of mechanical power*
- (1) Brick and tile presses
- (2) Opening or teasing machines in upholstery or bedding works
- (3) Carding machines in wool textile trades
- (4) Corner staying machines
- (5) Dough brakes
- (6) Dough mixers
- (7) Worm pressure extruding machines
- (8) Gill boxes in wool textile trades
- (9) The following machines used in laundries
  - (i) hydro-extractors
  - (ii) calenders
  - (iii) washing machines
  - (iv) garment presses
- (10) Meat mincing machines
- (11) Milling machines in use in metal trades
- (12) Pie and tart making machines
- (13) Power presses, including hydrolic and pneumatic presses
- (14) Loose knife punching machines
- (15) Wire stitching machines
- (16) Semi-automatic wood turning lathes

(b) *Machines whether or not worked with mechanical power*
(17) Guillotine machines
(18) Platen printing machines

Hoists and lifts (section 22)

4.68 Every hoist or lift shall be of good mechanical construction, of sound material and adequate strength, and shall be properly maintained. They must be examined thoroughly by a competent person at least every six months, and a report entered into or attached to the general register. If such examination reveals that certain repairs need to be carried out, a copy of the report must be sent to the HSE inspector. There must be a substantial enclosure and outer gates so that when the gates are shut, it is not possible for any person to fall down the liftway or to come into contact with any moving part of the hoist or lift. There must be an efficient interlocking or other device to ensure that the gate cannot be opened except when the cage or platform is at a landing, and the cage or platform cannot move away from the landing until the gate is closed. The maximum load which can be safely carried shall be marked conspicuously, and it is an offence to carry a greater load. The provisions as to gates, etc do not apply to continuous lifts or hoists (i e pater nostas) nor to ones which do not have mechanical power.

4.69 It will be noted that the section imposes an absolute duty. Good mechanical construction, sound materials, and adequate strength must exist not only at the time of purchase,-but throughout the working life of the lift or hoist. 'Sound materials' means materials which are sound, and not which appear to be sound (*Whitehead v James Stott & Co Ltd*). 'Maintained', according to section 176, means 'maintained in an efficient state, in efficient working order, and in good repair'.

Hoists and lifts for carrying persons (section 23)

4.70 Additional requirements are laid down when hoists and lifts are used for carrying persons, whether with goods or otherwise. Efficient and automatic devices shall be provided and maintained to prevent the cage or platform from overrunning, every point of access shall have a gate with an efficient device to ensure that the cage cannot be raised or lowered unless the gate is closed, and the cage will come to rest when the gate is open.

Teagle openings (section 24)

4.71 Every teagle opening or similar doorway used for hoisting or lowering goods or materials shall be securely fenced and shall be provided with a handrail on each side. The fencing shall be properly maintained and, except when the hoisting or lowering of goods or materials is being carried on, shall be kept in position.

Exceptions (section 25)

4.72 A lifting machine is not to be regarded as being a hoist or lift within the meaning of the above sections unless it has a platform or cage the direction of movement of which is restricted by a guide or guides. Thus a fork lift truck is not a hoist or lift (*Oldfield v Reed and Smith Ltd*). The power of the Secretary of State to grant exemptions from the requirements of sections 22–25 has been repealed, but the Hoists Exemptions Order 1962 (which is still in force) sets out a list of classes or descriptions of hoists and hoistways which come within the exemption Order, specifying the requirements of the Act which do not apply, subject to the conditions laid down in the Order.

Chains, ropes and lifting tackle (section 26)

4.73 No chain, rope or lifting tackle used for the purpose of raising or lowering persons, goods or materials, shall be used unless it is of good construction, sound materials, adequate strength and free from patent defect. A defect is 'patent' when it is visible, even though no-one notices it (*Sanderson v National Coal Board*). Further, the words 'patent defect' are not qualifying words, and do not weaken the statutory obligation so far as latent defects are concerned. In other words, they impose an additional requirement, not a lowering of standards. A table showing the safe working loads shall be posted in the stores where they are kept, and in a prominent position on the premises. Chains, ropes or lifting tackle must not be used in excess of the safe working load as shown on the table, they must be examined by a competent person at least every six months, and (except for fibre rope or fibre rope slings) must not be used in a factory for the first time unless they have been thoroughly examined and tested by a competent person, and a certificate stating the safe working load has been obtained and kept available for inspection. Relevant particulars must be entered into the register (see Chains, Ropes and Lifting Tackle (Register) Order 1938).

4.74 The purpose of section 26 is to prevent persons from being injured, either from a fall whilst they are being lifted, or from some object falling from broken lifting tackle, which must have physical propensities adequate for the job. The section does not lay down any requirement that the tackle shall be suitable for the purpose. Thus in *Beadsley v United Steel Companies Ltd* an accident resulted because the wrong type of lifting tackle was selected. The tackle was sound and of good construction, but was unsuitable for the task in hand. It was held that there was no breach of section 26. However, a civil claim based on negligence would probably succeed in such circumstances (*Dawson v Murex Ltd*).

Cranes and other lifting machines (section 27)

4.75 A lifting machine is any crane, crab, winch, teagle, pulley, block, gin wheel, transporter or runway. It has been held that a fork lift truck is a lifting machine (*McKendrick v Mitchell Swire*). All parts and working gear, whether fixed or moveable, including the anchoring and fixing appliances, of every lifting machine shall be of good construction, sound material, adequate strength, free from patent defect, and shall be properly maintained. Again, 'patent defect' is an additional requirement, for latent defects are clearly within the scope of the section (*McNeil v Dickson and Mann Ltd*). It has been held that an electrical operating and control button as within the meaning of 'parts and working gear', so a workman who injured his finger while pressing a stiff button was entitled to bring a successful claim under the Act (*Evans v Sanderson Bros and Newbould Ltd*), though the decision must be a marginal one.

4.76 All parts and gear shall be thoroughly examined by a competent person at least once in every period of fourteen months, and particulars entered into a register. If the examination shows that the lifting machine cannot be used safely without certain repairs being carried out, a copy of a report to that effect shall be sent to the HSE inspector (see Lifting Machines (Particulars of Examinations) Order 1963). All rails, tracks, etc shall be of proper size and adequate strength, and have an even running surface. They shall be properly laid, adequately supported or suspended, and properly maintained.

4.77 Every lifting machine shall be marked with its safe working load, except a jib crane with a variable working load which depends on the raising or lowering of the jib, in which case there must be an automatic indicator of safe working loads, or a table indicating the safe working load at the corresponding inclination of the jib or corresponding radii of the load. A lifting machine shall not be used for the first time in a factory unless it has been tested and thoroughly examined by a competent person and a certificate specifying the safe working load has been signed by him and kept available for inspection. Thereafter, no lifting machine shall be loaded beyond the safe working load except for the purpose of a test.

4.78 If any person is employed or working on or near the wheel track of an overhead travelling crane in any place where he would be liable to be struck by the crane, effective measures shall be taken by warning the driver of the crane to ensure that the crane does not approach within 20 feet (6 metres) of that place. 'Any place' does not mean a fixed spot, but a broad area (*Holmes v Hadfields Ltd*). The measures taken must be effective; thus, if an accident occurs, the measures can hardly be said to be effective. In *Lotinga v North Eastern Marine Engineering Co (1938) Ltd*, a workman was killed whilst working near the wheel track of an overhead crane. A notice was affixed to the crane that it must not be allowed to approach within 20 feet of men working. This was held not to be effective measures.

4.79 If a person is working above floor level, where he would be liable to be struck by an overhead travelling crane, or by a load carried by such crane, effective warnings must be given, unless his work is so connected with or dependent on the movement of the crane as to make a warning unnecessary.

Floors, passages and stairs (section 28(1))

4.80 All floors, steps, stairs, passages and gangways shall be of sound construction, properly maintained, and shall, so far as is reasonably practicable, be kept free from any obstruction and from any substance likely to cause persons to slip.

4.81 The opening words of the section must be interpreted in accordance with a certain amount of common sense. Thus a roadway (*Thornton v Fisher Ludlow Ltd*) or 'mother earth' (*Newberry v Westwood & Co*), or planks laid in an irregular manner over a gantry (*Tate v Swan Hunter and Wigham Richardson Ltd*) do not constitute 'floors', but planks across a duct may be a 'gangway' (*Hosking v De Havilland Aircraft Co Ltd*).

4.82 The floors must be of sound construction, which means fit for the work which it is anticipated is to be done on it (*Mayne v Johnstone and Cumbers Ltd*). Thus if a heavy machine or object falls on to the floor, causing it to give way, it may be that the floor is still sound for normal purposes, as long as it can withstand such further stress as may reasonably be expected to occur (*Mayne v Johnstone and Cumbers Ltd*). Sound construction means well made, not well designed, but if the floor becomes worn through normal use, it may well be that it has not been properly maintained (*Fisher v Port of London Authority*).

4.83 An obstruction is something which should not be there. Thus if bales, packages, etc are being stored on the floor, these do not constitute obstructions (*Pengelley v Bell Punch Co Ltd*). A trolley being used in the ordinary course of work is not an obstruction (*Marshall v Ericsson Telephones*) but it may be so if it was left unused for an unnecessary period of time.

4.84 Whether a substance is likely to cause someone to slip is probably a question of foreseeability. Thus water (*Taylor v Gestetner Ltd*), oil (*Latimer v AEC Ltd*), and grease (*Williams v Painter Bros Ltd*) are obvious examples. In *Dorman Long (Steel) Ltd v Bell*, metal plates left temporarily on the floor became slippery when slag dust collected on them, and it was held that it was irrelevant that the slippery substance was not actually in contact with the floor.

4.85 It will be noted that as far as the requirements which relate to obstructions and slippery substances are concerned, they are qualified by the words 'so far as is reasonably practicable'. This again must be a question of fact; in some circumstances it may not be possible to remove the offending items immediately (*Jenkins v Allied Ironfounders Ltd*), or the employers may be able to point to a regular system of cleaning up the floors or removing obstructions (*Braham v J Lyons & Co*). As we have noted, the burden of proof is on the defendants to show that it was not reasonably practicable to do more than was done. In *Bennett v Rylands Whitecross Ltd* the plaintiff tripped over a piece of wire and was injured. There was no explanation as to how the wire got there, and it was held that the fact that the obstruction might have been created unwittingly or in a highly improbable manner did not absolve the defendants from liability unless they could show that it was not reasonably practicable for them to have prevented or removed the obstruction.

Staircases, rails (section 28(2), (3))

4.86 Every staircase in a building or forming the means of exit shall have a substantial handrail provided and maintained, and if there is an open side, it shall be on that side. If there are two open sides, there must be a handrail on each side. To be substantial, the handrail must act as a guardrail, as well as for holding purposes (*Corn v Weir's Glass (Hanley) Ltd*). The Act refers to a staircase, not to stairs per se. In *Kimpton v Steel Co of Wales Ltd* it was held that three steps leading to a machine did not constitute a staircase. Any open side of a staircase must also have a lower rail or other effective guard (section 28(3)).

Openings in floors (section 28(4))

4.87 All openings in floors shall be securely fenced, except when the nature of the work renders such fencing impracticable. In *Barrington v Kent Rivers Catchment Board* the plaintiff fell into an open inspection pit at a garage. At the time, no work was being carried on, and thus the employers could not successfully argue that the nature of the work rendered it impracticable to fence. The fact that it may be inconvenient to fence does not make it impracticable to do so (*Street v British Electricity Authority*).

Ladders (section 28(5))

4.88 All ladders shall be soundly constructed and properly maintained. Again, the duty is an absolute one, and it is no excuse that the employer took all reasonable steps to ensure that ladders were of sound construction (*Cole v Blackstone & Co Ltd*).

Safe means of access (section 29(1))

4.89 So far as is reasonably practicable, there shall be provided and maintained a safe means of access to every place where any person has to work. The means of access must be safe wherever they lead to. In *Lavender v Diamints Ltd* a window cleaner climbed over a roof made from asbestos sheeting in order to reach some windows. He fell through the roof and was injured. It was held that the occupiers were liable. The subsection was not confined to the means of access to the inside of a factory, but to any place where a person had to work. For example, if a person has to climb a ladder to get to his place of work, a failure to foot it at the bottom or to secure it at the top may mean that the means of access are not safe (*Geddes v United Wires Ltd*). But the access must be to a place of work, not, for example, to a canteen (*Davies v De Havilland Aircraft Co Ltd*) or to a lavatory (*Rose v Colvilles Ltd*).

4.90 The means of access may be unsafe because of some structural defect, or obstructions, or defect in the floors, etc. Since the subsection contains the words 'so far as is reasonably practicable', there is no guarantee of safety, especially if there is a misuse of the means of access by others which cause the injury. For example, in *Higgins v Lyons & Co Ltd* the plaintiff was injured by a lorry which was being driven in a yard leading to the place of work, and the employers were held not liable. It has been suggested that the obligation is to provide and maintain a means of access which is structurally sound, and that this does not impose obligations to eliminate transient and temporary conditions (*Levesley v Thomas Firth and John Brown Ltd*). Thus if the means of access becomes hazardous because of snow or ice, it may not be reasonably practicable to take immediate steps to do something about it, and a temporary delay may be expected (*Latimer v AEC Ltd*). However, there must be some reasonable attempts to deal with the problem (*Thomas v Bristol Aeroplane Co Ltd*).

4.91 The duty to provide a safe means of access appears to include not only the physical propensities of that access, but also the environmental conditions. Thus, if the means of access are unsafe because of excessive noise or heat, or insufficient lighting, etc there will be a breach of the section (*Carragher v Singer Manufacturing Ltd*). It also encompasses the risk of falling from heights (*Nimmo v Alexander Cowan Ltd*). However, the duty on the occupier is to provide safe means of access; if he does this, but workmen choose another route and are injured, the occupier may well escape liability (*Street v British Electricity Authority*). It should be noted that the duty of the occupier is to any person working on the premises, including an independent contractor and his employees (*Lavender v Diamints Ltd*).

Place of work (section 29(1))

4.92 So far as is reasonably practicable, every place of work shall be made and kept safe for any person working there. This refers to the actual working place, as opposed to the access to that place. However, similar considerations will apply. Thus if the place of work is safe to start with, but becomes unsafe because of some temporary or transient condition, it may be that there is no breach of the section, although there would be if the danger was permanent, or if it was reasonably foreseeable (*Evans v Sant*). Again, the duty is owed to every person who is working there, including an independent contractor, and it therefore exists irrespective of whether the independent contractor is under a separate duty, e g under the Construction Regulations (*Whincup v Woodhead & Sons Ltd*). It follows that if an independent contractor is carrying on work within a factory, the occupier must take some steps to familiarise himself with the situation and take reasonably practicable steps to deal with any potential hazard which may arise (*Taylor v Coalite Oils and Chemicals Ltd*).

Working at heights (section 29(2))

4.93 Where any person has to work at a place from which he will be liable to fall a distance of more than 6 feet 6 inches then, unless the place is one which affords a secure foothold (and, where necessary, a secure handhold), means shall be provided, so far as is reasonably practicable, by fencing or otherwise, for ensuring his safety. A handhold is something which a person can hold onto from time to time, e g the upright of a ladder (*Wigley v British Vinegars Ltd*). A foothold does not cease to be secure because in abnormal circumstances (such as an explosion) a person is thrown off (*Tinto v Stewarts and Lloyds Ltd*). The burden of proof is on the plaintiff to show that the absence of a secure foothold or handhold was responsible for the accident.

4.94 If the occupier is unable to provide a secure foothold or handhold, other safety precautions must be provided. This can be fencing, the provision of safety belts (*McWilliams v Sir William Arrol & Co Ltd*) or other suitable means.

Confined spaces (section 30)

4.95 If work has to be done in any chamber, tank, vat, pit, pipe, flue or similar confined space in which dangerous fumes are likely to be present to such an extent as to involve a risk of persons being overcome by them, certain precautions must be observed. Unless there is some other adequate means of exit, there must be a manhole, not less than 18 × 16 inches (460 × 410 millimetres) (or, if circular, 18 inches (460 millimetres) in diameter) or, in the case of tank wagons or other mobile plant, 16 × 14 inches (410 × 360 millimetres) (or 16 inches (410 millimetres) in diameter if circular). No person shall enter or remain therein unless he is wearing suitable breathing apparatus, has been authorised to enter by a responsible person, and, where practicable, is wearing a belt with a rope securely attached. There must be a person keeping watch outside and capable of hauling him out by holding the free end of the rope. Breathing apparatus need not be used if the confined space has been certified to be free from fumes for a specified period by a responsible person, as long as the person entering the confined space has been warned when the safe period will expire. A sufficient supply of approved breathing apparatus, of belts and ropes, and of reviving apparatus and oxygen, shall be provided and kept readily available, and shall be thoroughly examined at least once a month, with a report, signed by a competent person, kept available for inspection. A sufficient number of persons shall be trained and have adequate practice in the use of the apparatus and in the methods of restoring respiration (see the Examination of Breathing Apparatus Order 1061).

4.96 No person shall enter or remain in any confined space in which the oxygen in the air is liable to have been substantially reduced unless he is wearing breathing apparatus, or the space has been tested and certified as being safe for entry with such apparatus. No work shall be permitted in any boiler, furnace or boiler flue until it has been sufficiently cooled so as to make it safe.

Explosive or inflammable substances (section 31)

4.97 Where, in connection with any grinding, sieving or other process giving rise to dust which may cause an explosion on ignition, all practicable steps shall be taken to prevent such explosion by enclosure of the plant, by removal or prevention of accumulation of dust, and by the exclusion or effective enclosure of possible sources of ignition. This section is designed to deal with the dangers inherent in the grinding of coal or other carbonaceous material and other dust likely to explode if ignited. Unless the plant is so constructed as to withstand the pressure likely to be produced from any such explosion, all practicable steps shall be taken to restrict the spread and effects by the provision of chokes, baffles, vents or other equally effective appliances. Special provisions must be taken in connection with the grinding of magnesium (Magnesium (Grinding of Castings) Special Regulations 1946).

4.98 Before any part of a plant which contains explosive or inflammable gas or vapour under pressure is opened, the flow shall be effectively stopped by a stop value or otherwise, and before the fastening is removed, the gas or pressure must be reduced to atmospheric pressure.

4.99 Any tank plant or vessel which has contained any explosive or inflammable substance shall not be subjected to any welding, brazing or soldering operation, or any cutting operation, which involves the application of heat for the purpose of taking it apart unless all practicable steps have been taken to remove the substance, or to render it non-explosive or non-inflammable. Once such operation has been carried out, no explosive or inflammable substance shall be allowed to get into the relevant container until the metal has cooled sufficiently to prevent the risk of ignition taking place. HSE has power to make certain exemptions from these latter provisions.

Steam boilers (sections 32–34)

4.100 Every part of every steam boiler shall be of good construction, sound material and adequate strength, and free from patent defect. Attached to steam boilers shall be a suitable safety valve, stop-value, correct steam pressure gauge, at least one water gauge, sufficient means for attaching a pressure gauge, and, unless externally fired, shall be provided with a suitable fusible plug or an efficient low water alarm device. There are a number of exemption certificates in force in respect of particular types of steam boilers and those manufactured by certain firms.

4.101 Every steam boiler and all its fittings and attachments shall be properly maintained. It shall not be used in a factory unless it has been examined in accordance with the Examination of Steam Boiler Regulations 1964. A report of every such examination shall be entered into or attached to the general register signed by the person making the examination. A new steam boiler shall not be used unless the manufacturer has issued a certificate specifying the maximum permissible working pressure (section 33, and see the Examination of Steam Boilers Reports (No 1) Order 1964). Again, a number of exemption certificates are in force.

4.102 No person shall enter into any steam boiler which is one of a range of two or more unless all inlets through which steam or hot water might enter have been disconnected, or all valves controlling the entry of steam or hot water are closed and securely locked.

Steam receivers and steam containers (section 35)

4.103 Every part of every steam receiver shall be of good construction, sound material, adequate strength and free from patent defect. Steam receivers shall be properly maintained, and thoroughly examined by a competent person at least once in every twenty-six months. A report containing the prescribed particulars shall be entered into or attached to the general register. Again, a number of exemption certificates are in force.

4.104 Every steam receiver which is not constructed so as to withstand with safety the maximum pressure which can be obtained from any other source of supply shall be fitted with

(a) a suitable reducing valve or other automatic device to prevent the safe working pressure being exceeded;
(b) a suitable safety valve so adjusted as to permit the steam to escape as soon as the safe working pressure is exceeded;
(c) a correct steam pressure gauge;
(d) a suitable stop valve;
(e) a plate bearing a distinctive number which is easily visible (except when only one steam receiver is in use).

Air receivers (section 36)

4.105 Every air receiver and its fittings shall be of sound construction and properly maintained. They shall have marked on them the safe working pressure, and if connected to an air compressor plant, shall be able to withstand the maximum pressure or be fitted with a suitable reducing valve. A suitable safety valve shall be fitted, so adjusted as to permit air to escape as soon as the safe working pressure is exceeded, and shall be fitted with a correct pressure gauge, a draining appliance, means whereby the interior can be cleaned, and, if more than one is in use in the factory, a distinguishing mark which is easily visible. Every air receiver shall be thoroughly cleaned and examined at least once in every twenty-six months except in the case of a receiver of solid drawn construction which can be examined within four years from the previous examination. The examination and test shall be carried out by a competent person, and a report, containing the prescribed particulars shall be attached to or entered in the general register. Again, a number of exemptions certificates are in force.

4.106 It should be noted that the provisions relating to steam boilers and air receivers are specifically stated to apply to building operations and to works of engineering construction undertaken by way of trade or business, or for the purpose of any industrial or commercial undertaking (see section 127). However, sections 32–34 (above) do not apply to boilers belonging to or exclusively used by the Crown, or belonging to and used by the UK Atomic Energy Authority, or belonging to and used by ships or railway companies.

Water sealed gasholders (section 39)

4.107 Every gasholder which has a storage capacity of at least 5,000 cubic feet (140 cubic metres) shall be of sound construction and properly maintained. They must be examined externally by a competent person at least every two years, and a record containing the prescribed particulars entered into or attached to the general register (see the Gasholders (Record of Examinations) Order 1938).

General welfare provisions

4.108 Part III of the Act is designed to ensure that certain minimum standards of welfare are observed. The fact that the employees are contented with existing facilities, or have not complained, is irrelevant (*File Tile Distributors Ltd v Mitchell*, see chapter 2). At one time it was suggested that no civil action can be based on a breach of these provisions, although claims succeeded in those cases where the point does not appear to have been taken, and it is now believed that the weight of authority is in favour of such claims. It will be noted that the provisions are for the benefit of persons employed.

Supply of drinking water (section 57)

4.109 An adequate supply of wholesome drinking water shall be provided and maintained at suitable points conveniently accessible to all persons employed. If the supply is not laid on, it shall be contained in suitable vessels, renewed daily, and all practicable steps shall be taken to preserve the water and the vessels from contamination. Except where there is an upward jet, one or more suitable cups shall be provided at each point of supply.

Washing facilities (section 58)

4.110 There shall also be provided and maintained for use of employed persons adequate and suitable facilities for washing, including a supply of clean running hot and cold water, soap and clean towels or other suitable means for drying, and these facilities shall be conveniently accessible and shall be kept clean and in orderly condition. In *Reid v Westfield Paper Co Ltd* it was held that an employee who suffered dermatitis as a result of the employer's failure to provide adequate washing facilities could recover damages for that breach. An HSE inspector may issue an exemption certificate in certain cases (see Washing Facilities (Running Water) Exemption Regulations 1960).

Accommodation for clothing (section 59)

4.111 There shall be provided and maintained for the use of employed persons adequate and suitable accommodation for clothing not worn during working hours, and such arrangements as are reasonably practicable for drying such clothing. In *McCarthy v Daily Mirror* it was held that in determining whether or not the accommodation provided was suitable, regard must be had to the possibility of the clothing being stolen, so that if it is unsuitable by this standard, an employee may be able to bring a successful claim against the employer.

Sitting facilities (section 60)

4.112 Where it is possible for employees to have reasonable opportunities to sit in the course of their work, suitable and sufficient facilities shall be provided and maintained. Where a substantial proportion of the work can be done seated, there shall be provided a seat of a design, construction and dimensions suitable for the employee and for the work, together with a foot rest, and the seat shall be properly supported while in use for that purpose.

First aid (section 61)

4.113 This section (and Regulations made thereunder) has been repealed by the Health and Safety (First Aid) Regulations 1981, which will come into force in July 1982. An Approved Code of Practice and Guidance Notes have also been issued (see chapter 6).

Removal of dust or fumes (section 63)

4.114 If a process is carried on which

(a) gives off any dust or fumes or other impurity of such a character and to such an extent as to be likely to be injurious or offensive to persons employed, or

(b) any substantial quantity of dust of any kind,

then all practicable measures shall be taken to prevent persons employed from inhaling the dust or fume or other impurity, and to prevent it accumulating in any workroom. Where the nature of the process makes it practicable, exhaust appliances shall be provided and maintained as near as possible to the point of origin of the dust or fume or other impurity, so as to prevent it entering the air of any workroom.

4.115 It will be noted that this section has two parts. The first deals with dust, fume or other impurities as are likely to be injurious or offensive. Whether this is so can only be determined by the knowledge which the factory occupier has or ought reasonably to have. Thus, he will not be liable under the section if the danger is unknown, for then it is not possible for him to take all practicable steps (*Ebbs v James Whitson & Co Ltd*), but he will be liable if he ought to know of the danger, or if he fails to keep up to date with current knowledge and developments. Nor can he be liable for failing to use a safety device which was not invented at the time of the incident (*Adsett v K and L Steelfounders and Engineers Ltd*). Section 63 is concerned with protective measures (as opposed to section 4, above, which deals with ventilation). If, therefore, masks would provide an adequate protection against the dust, etc, the employer must take all practicable measures, which means that not only must he provide masks, but take steps to ensure that the employees use them (*Crookall v Vickers Armstrong Ltd*), for the dangers are frequently insidious, and the employees would not readily appreciate the risks involved.

4.116 The second part to this section comes into operation where any substantial quantity of dust of any kind is given off. This prohibition is absolute in its terms, and is not dependent on whether or not it is likely to be injurious. Nonetheless, the occupier need only take all practicable steps, which is a standard slightly less than being absolute, and which must depend on the current knowledge of the danger at the time of the alleged breach (*Richards v Highway Ironfounders Ltd*).

4.117 Additionally, there are a number of special Regulations which deal with particular processes, and which may have some effect in modifying or superseding the provisions of section 63. These include

(a) Chemical Works Regulations 1922;
(b) Grinding of Metals (Miscellaneous Industries) Regulations 1925 and 1950;
(c) Pottery (Health and Welfare) Special Regulations 1950;
(d) Asbestos Regulations 1969;
(e) Abrasive Wheels Regulations 1970.

4.118 So far as a civil action for damages is concerned, the plaintiff must prove that on the balance of probabilities the breach of the statutory duty caused the injury, and that if proper protection had been given or provided, he would have used it (*Richards v Highway Ironfounders Ltd*).

Meals in dangerous trades (section 64)

4.119 Where in any room, arsenic or any other poisonous substance is so used as to give rise to any dust or fume, a person shall not be permitted to partake of food or drink in that room or to remain there during intervals allowed to him for meals or rest (other than intervals allowed in the course of a spell of continuous employment). Suitable provision shall be made for persons employed in any such room to take their meals elsewhere in the factory. For similar provisions relating to lead, see Control of Lead at Work Regulations 1980 (chapter 6).

Protection of eyes (section 65)

4.120 In any process which involves a special risk of injury to the eyes from particles or fragments thrown off in the course of the process, the Secretary of State may make regulations requiring suitable goggles or effective screens to be provided. This section is for the benefit of the employees of the factory occupiers (*Whalley v Briggs Motor Bodies Ltd*). Eye protectors must be suitable for the work and the individual, and must be adapted for the work to be done (*Daniels v Ford Motor Co*). The obligation is to provide; this means that the employee must be informed that they are available, and told from where they can be obtained (*Finch v Telegraph Construction and Maintenance Co Ltd*), which must be reasonably at hand, but there is no absolute obligation to compel their use (*Norris v Syndic Manufacturing Co Ltd*).

4.121 The Protection of Eyes Regulations 1974 (as amended) specify certain processes for which an employer must provide eye protection and require that persons engaged in those processes or at risk from such processes wear or use the protection provided. The protection provided must be of a type approved by certificate of an authorised inspector (generally, this means conforming to British Standards) and may consist of goggles, visors, spectacles and face screens or fixed shields, as appropriate. The Regulations apply to all factories, and to electrical stations, charitable and reformatory institutions, certain dock premises and warehouses, ships and works of building and engineering construction, Exemption certificates may be granted if the inspector considers that the requirements are not necessary for the protection of persons employed or that it is not reasonably practicable to use them.

4.122 Eye protectors shall be given by the employer into the possession of the person for whom they are provided except in the case of persons who are only occasionally employed. In the latter case, a sufficient number of eye protectors shall be provided, maintained, and kept readily available by their employer. If eye protectors are lost or destroyed or become defective, replacements must be handed to the persons for whom they are provided, and the employer shall keep available as many eye protectors as will be sufficient, so far as can be reasonably foreseen, to comply with the provisions of the Regulations.

4.123 Eye protectors and shields shall be suitable for the person for whose use they are provided. Thus, for a person who already wears spectacles, special prescription spectacles with ophthalmic lenses must be provided, otherwise plain lenses will presumably suffice. They must be made in accordance with the approved specification, and marked accordingly. Fixed shields must also conform to approved specifications, be properly maintained, and if it is necessary that they should be transparent, be kept clean, and so constructed and kept in position as to protect, so far as practicable, the eyes of persons for whose protection they are provided while those persons are carrying on a specified process, or while they are employed in a place where there is a reasonably foreseeable risk of injury to their eyes from such processes.

4.124 Every person who is provided with eye protectors or a shield shall use them while he is employed in a specified process or while he is employed in a place where there is a reasonably foreseeable risk of injury to his eyes from the carrying on of the process, unless a fixed shield is provided. He shall take reasonable care of the eye protectors or shield and not wilfully misuse them, and report forthwith to the employer (or his agent) the loss or destruction of them, or any defect in them. He shall also make proper use of them.

4.125 Schedule 1 of the Regulations requires the employer to provide the protection which is appropriate for his employees. Schedule 2 lays down the precautions to be provided for the benefit of persons who are not employed on Schedule 1 processes, but who are employed in a place where there is a reasonably foreseeable risk of injury to their eyes from the carrying on of the process. The full requirements can be found in the Appendix to this book.

4.126 Other Regulations which require the provision of eye protectors are noted in chapter 6.

Miscellaneous premises (sections 68–69)

4.127 These sections deal with humid factories and underground rooms respectively.

Lifting of heavy weights (section 72)

4.128 A person shall not be employed to lift, carry or move any load so heavy as to be likely to cause injury to him. If a person has been properly trained and is experienced, the fact that the load is a heavy one does not mean that there is a breach of this section (*Kinsella v Harris Lebus Ltd*). If the weight is too heavy to be moved unaided, assistance must be provided, and it is immaterial that the employee could have had assistance had he asked for it (*Brown v Allied Ironfounders Ltd*). However, if there are a number of persons employed to carry a heavy weight, the fact that the weight is too much for one to carry on his own is irrelevant as long as it is within the collective capability of them all. There are special provisions which lay down the maximum weights which can be lifted by men, women and young persons respectively in the

Woollen and Worsted Textiles (Lifting of Heavy Weights) Regulations 1926, Jute (Safety, Health and Welfare) Regulations 1948, and Pottery (Health and Welfare) Special Regulations 1950 (see chapter 7).

Female young persons (section 73)

4.129 A young female person (i e a girl under the age of eighteen) shall not be employed in any part of a factory where certain glass-making processes, and certain activities involving brine or salt, are being carried on.

Women and young persons in certain lead processes (sections 74, 131)

4.130 Women and young persons shall not be employed in certain lead processes being carried on in a factory, and may not be employed in painting any part of a building with lead paint, other than as apprentices or doing certain decorative work. Other restrictions on women and young persons working with lead are now to be found in the Control of Lead at Work Regulations (see chapter 6).

Notification of industrial diseases (section 82)

4.131 If a medical practitioner believes that a patient of his is suffering from lead, phosphorus, arsenical or mecurial poisoning, or anthrax, which may have been contracted in any factory, the doctor shall forthwith send details to HSE. Similar details must be sent by the occupier of the factory. The Factories (Notification of Diseases) Regulations 1966 adds to the list of notifiable diseases any incidence of acute, sub-acute or chronic diseases of any organ due to berylium, cadmium, lead, arsenic, mercury, certain phosphates, and their compounds or alloys.

Hours of work of women and young persons

4.132 Part VI of the Factories Act (together with the Hours of Employment (Conventions) Act 1936) imposes certain restrictions on the hours of work of women (over the age of eighteen) and young persons (of either sex) who are above compulsory school-leaving age but who are below the age of eighteen. A review of all these provisions, to ascertain whether they are necessary in modern conditions, has been undertaken by the Equal Opportunities Commission, and, together with HSC, further consideration is being given to the elimination of a protective legislation in so far as it involves the separate treatment of women.

4.133 The main statutory provisions relating to factories are as follows:

Hours of work (section 86)

4.134 The total number of hours worked by women and young persons shall not exceed nine in any one day, nor forty-eight in any week. The time for commencing employment shall not be earlier than 7 a.m. and they must finish before 8 p.m. (1 p.m. on Saturdays). The actual spell of work may not exceed four-and-a-half hours without there being a break of at least half an hour, or five hours if there has been a break of at least ten minutes. The times of employment, and the intervals, must be the same for all women and young persons, and they may not be employed during the intervals allowed for meals or rest.

Notice fixing hours (section 88)

4.135 A notice, fixing the hours and intervals for meals and rest, must be posted in the factory, and no woman or young person may be employed otherwise than at the times specified.

Overtime (sections 89–90)

4.136 Women and young persons may work overtime in addition to the above limits, provided that the total hours worked do not exceed ten in any one day, the period of employment does not exceed twelve hours, and does not start before 7 a.m. or finish after 9 p.m. (1 p.m. on Saturdays). Not more than six hours overtime may be worked in any one week, nor more than 100 hours overtime in any one year, and the overtime must not be in more than twenty-five weeks in the year.

4.137 The Secretary of State may, by Regulation, increase the permissible hours of overtime in the case of factories which are subject to seasonal work. The following Regulations have been made:

(a) Areated Water Manufacture (Overtime) Regulations 1938;
(b) Biscuit Manufacture (Overtime) Regulations 1938;
(c) Bottling of Beer, Wines and Spirits (Overtime) Regulations 1940;
(d) Chocolate and Sugar Confectionery (Overtime) Regulations 1939;
(e) Dyeing and Cleaning (Overtime) Regulations 1939;
(f) Florists (Overtime) Regulations 1938;
(g) Glass Bottles and Jars (Overtime) Regulations 1938;
(h) Laundries (Overtime) Regulations 1938;
(i) Net Mending (Overtime) Regulations 1939;
(j) Poultry Preparation (Overtime) Regulations 1958;
(k) Factory Overtime (Separation of Sets) Regulations 1938;
(l) Factory (Individual Overtime) Regulations 1938.

4.138 Before employing any woman or young person on overtime, the factory occupier shall send to the HSE inspector such particulars as may be prescribed, and enter details in the register. A notice containing these particulars must also be posted in the factory.

Employment during intervals (sections 91–92)

4.139 Women and young persons may not be employed outside the factory on any business carried on by the occupier during the periods allowed for meals or rest, nor may they be allowed to remain in a room where processes are being carried on during these periods.

Sunday employment (section 93)

4.140 Women and young persons may not be employed on Sunday in a factory or on any other business carried on by the occupier.

Exceptions (sections 95–115)

4.141 There are a number of exceptions to the above provisions. These include:

(a) women who hold responsible positions of management and who are not ordinarily engaged in manual work (section 95);
(b) where HSE suspends the provisions in the event of accident, breakdown of machinery or plant, or other unforeseen emergency (section 96);
(c) where HSE authorises a double day shift system (section 97, and see Shift System in Factories and Workshops (Consultation of Workpeople) Order 1936);
(d) employment of male young persons in shifts in certain industries (section 99, and see Employment of Young Persons (Iron and Steel Industry) Regulations 1959, Employment of Young Persons (Glass Containers) Regulations 1955, Night Work of Male Young Persons (Medical Examinations Regulations 1938);
(e) where a five day week is being worked (section 100);
(f) where the exigencies of the trade or the convenience of the employees so require (section 101, and see Bread, Flour Confectionery and Sausage Manufacture (Commencement of Employment) Regulations 1939);
(g) certain continuous processes (section 102, and see Factories (Intervals for Women and Young Persons) Regulations 1938);
(h) certain processes (section 103);
(i) young male persons employed in continuous employment with men (section 105);
(j) young male persons employed on repair work (section 106);
(k) certain factories which work on Saturdays (section 107, and see Factories (Saturday Exception) Regulations 1940);
(l) annual holidays on different days (section 108, and see Factories Act Holidays (Different Days for Different Sets) Regulations 1947);
(m) factories occupied by members of the Jewish faith (section 109);
(n) laundries (section 110, and see Laundries, Manufacture of Bread etc (Hours and Intervals) Modification Regulations 1938);
(o) bread, flour confectionery and sausage manufacturing (section 111);
(p) preserving of fish, fruit or vegetables (section 112, and see Fruit and Vegetable Preserving (Hours of Women and Young Persons) Regulations 1939);
(q) treatment of milk (section 113, and see Milk and Cheese Factories (Hours of Women and Young Persons) Regulations 1949).

4.142 Before an occupier can avail himself of any of these exceptions, he must serve a notice on the HSE inspector at least seven days in advance, and post a notice in the factory (section 115).

Exemptions (section 117)

4.143 HSE has power to grant exemptions from the above provisions (and from the provisions of the Hours of Employment (Conventions) Act, and the Employment of Women, Young Persons and Children Act 1920) whenever HSE thinks that it is appropriate to do so, and desirable in the public interest for the purpose of maintaining or increasing the efficiency of industry or transport. The exemption may be special, i e for particular persons, employment or premises, or general. Special exemptions will last for one year (but are renewable), but general exemptions are not so limited. The following general exemptions have been made:

(a) Cotton Factories (Length of Spell Exemption) Order 1947;
(b) Factories (Evening Employment) Order 1950;
(c) Railway Employment Exemption Regulations 1962.

Night work

4.144 The Hours of Employment (Conventions) Act 1936 prohibits the employment of women in industrial undertakings at night, i e between the hours of 10 p.m. and 5 a.m. The Act does not apply to women holding managerial positions who are not engaged in manual labour. As noted, this Act is subject to the power of HSE to make exemption orders.

Annual holidays (section 94)

4.145 Every woman and young person employed in a factory shall be entitled to have Christmas Day, Good Friday and all other days specified as Bank Holidays (Banking and Financial Dealings Act 1971). Another day may be substituted if the occupier posts a notice in the factory giving at least three weeks notice of the substitution.

Certificate of fitness (section 119)

4.146 If an inspector is of the opinion that the employment of a young person in a factory or a particular process is prejudicial to his health or to the health of other persons, he may serve a written notice to this effect on the occupier. The latter will not be able to continue to employ that young person in that place or on that process until the Employment Medical Adviser has examined him and certified that he is fit to work in the factory or on the process.

Notice of employment of young persons (section 119A)

4.147 When a factory occupier takes a young person into employment, he shall, within seven days, send specified particulars to the local careers office.

Special applications (sections 121–127)

4.148 These sections make special provisions for applying the requirements of the Act (with appropriate modifications) to parts of buildings let off as a separate factory, electrical stations, charitable or reformatory institutions, docks, wharves, warehouses, ships, building operations and works of engineering construction.

Homeworkers (section 133)

4.149 In certain prescribed cases, the factory occupier (or contractor employed by him) shall keep a list of the names and addresses of all outworkers directly employed by him, and send a copy to the inspector and the local council. Further details can be found in the Homeworkers Order 1911. New Regulations dealing with homeworkers are expected to be in force in the near future (see chapter 12).

Notices and returns (section 137)

4.150 Every person who occupies premises as a factory must, one month prior to going into occupation, send to the inspector a notice stating the name or title of the occupier, the postal address of the factory, the nature of the work, stating whether or not mechanical power is being used, the name of the local authority, and other prescribed particulars.

Abstracts of the Act (section 138)

4.151 At the principal entrances of a factory there shall be posted an abstract of the Act, a notice stating the address of the inspector, the address of the Employment Medical Adviser for the area, a notice specifying the clock (if any) by which periods of employment and rest periods are to be taken, and every other notice required to be posted. Printed copies of any special Regulations in force for the factory or a prescribed abstract shall be kept where they may be conveniently read.

General Register (section 140)

4.152 Every factory shall have a General Registrar (Form F31 is available for this purpose, obtainable from HMSO). It is in seven parts, and should contain the following information:

Part 1 – name of occupier and address of factory;
nature of the work carried on;
any exceptions in respect of the employment of women and young persons under sections 99–113;
names of appointed machinery attendants for the purpose of the Unfenced Machinery Regulations 1938 and 1946;
particulars of the fire certificate;
Part 2 – particulars of young persons employed;
Part 3 – particulars of accidents and dangerous occurrences;
Part 4 – particulars of prescribed cases of poisoning or diseases;
Part 5 – details of washing, whitewashing, or colour washing, or painting or varnishing of inside walls, partitions or ceilings;
Part 6 – particulars of the testing or examination of fire warning systems;
Part 7 – particulars of persons who are trained in first aid.

Attached to the General Register shall be kept all other matters which are required to so be, e g reports on the examination of hoists and lifts (sections 22–25) steam boilers, steam receivers and air receivers (sections 32–38) and other matters specified in the appropriate Regulations.

4.153 The General Register shall be kept for inspection by an inspector or employment medical advisers for at least two years after the date of the last entry.

Enforcement of the Act (section 155)

4.154 The main liability for criminal penalties is placed on the occupier of the factory, or, in certain cases, the owner. If a person contravenes the provisions of a Regulation or Order made under the Act, which expressly imposes a duty on him, then that person shall be guilty of an offence, and the owner or occupier shall not be guilty of an offence unless it is proved that he failed to take all reasonable steps to prevent the contravention (section 155(2)). The onus is on the prosecution to prove that the occupier or owner failed to take all reasonable steps to prevent the contravention (*Wright v Ford Motor Co Ltd*).

Penalties

4.155 The Factories Act is one of the existing statutory provisions for the purpose of section 33(3) of HSWA (see Schedule 1), and therefore offences are punishable by a fine of up to £1,000 on summary conviction, an unlimited fine in respect of a conviction on indictment, and, if the offence falls within section 33(4) of HSWA, a term of imprisonment of up to two years and/or a fine. These punishments do not apply to offences under sections 40–52, 135 and 146(4) of the Factories Act, for which section 156 of that Act still applies.

Application to the Crown

4.156 The Act applies to factories belonging to or in the occupation of the Crown, and to building operations or works of engineering construction undertaken by or on behalf of the Crown (section 173).

5 Offices, Shops and Railway Premises Act 1963

5.1 This Act applies to office premises, shop premises, and railway premises, being premises wherein persons are employed under a contract of employment (or apprenticeship) for an aggregate of at least twenty-one hours. Thus if only voluntary labour is working on the premises, or if casual workers are employed for less than twenty-one hours in total, the Act does not apply.

5.2 Also excluded are the following:

(a) any premises wherein the only person employed is the husband, wife, parent, grandparent, son, daughter, grandchild, brother or sister of the person by whom they are employed (section 2(1));
(b) premises of homeworkers (section 2(2));
(c) premises which form part of a factory for the purpose of the Factories Act, or which are below the ground and constitute a mine for the purpose of the Mines and Quarries Act (section 85). However, it is possible that part of the premises may be a factory and part a shop. In *Hoare v Green* (see chapter 4) a workroom behind a florists shop was held to be a factory, and thus covered by the provisions of the Factories Act, whereas the shop part of the premises where the flowers were sold would presumably come under the scope of OSRPA;
(d) premises to which exemption orders apply.

5.3 The Act does not confer on any member of certain visiting forces a right to sue his Government (or another member of those forces) in tort in respect of anything done or omitted to be done in the course of his duty. This applies to those Governments which are so specified under the Visiting Forces Act 1952 as well as organisations designated under the International Headquarters and Defence Organisations Act 1964 (section 84).

5.4 It is also a defence in any civil proceedings brought to recover damages for a breach of statutory duty (see chapter 9), or in any criminal proceedings based on a contravention of the Act, to prove that at the time of the breach or contravention the premises were being used for a temporary or transitional purpose of not more than six weeks in the case of a fixed structure, or six months in the case of a moveable structure (section 86). If the premises are to be used on a temporary basis, it is still necessary to notify the appropriate authority under section 49(1) (see para 5.58), although if a prosecution was brought in respect of a failure to do so, it would be a defence to show that the persons in question were employed in premises to be used only for temporary occupation (section 86(2)).

5.5 It will be recalled that the various statutory provisions only apply to those premises which fall within their respective scope, apart from HSWA, which is ubiquitous in nature, and applies to all employment scenes. But OSRPA applies only to those premises which fall exactly within the statutory definitions, and while there are many marginal situations, other exclusions are quite clear. Thus doctors' surgeries, night clubs, schools, dance halls, universities, etc, may all come under HSWA, but not other legislation unless there is a specific provision which applies.

5.6 Many of the provisions of OSRPA follow closely the pattern and wording which can be found in the Factories Act, and it is generally accepted that both statutes can be interpreted *in pari materia* (see para 1.59). For obvious reasons, more cases fall to be decided under the latter Act than the former.

Office premises (section 1(2))

5.7 These are defined as being a building, or part of a building, the sole or principal use of which is as an office or for office purposes. This latter phrase includes the purposes of administration, clerical work, handling money, and telephone and telegraph operating. Clerical work includes writing, bookkeeping, sorting papers, filing, typing, duplicating, machine calculating, drawing, and the editorial preparation of matter for publication. The term 'building' implies a permanent structure of some sort, and would therefore include a hut or shed etc, if used for office purposes. Any part of the premises which is ancillary to the main purpose of an office is also covered by the Act, e g canteens, dining-rooms, store-rooms, washrooms, lavatories (section 1(5)) although in respect of some of these rooms there are specific legal requirements.

Shop premises (section 1(3))

5.8 This phrase is slightly more complicated, and is defined as follows:

5.9 (a) A shop. Although this term was given a somewhat wide meaning in the Shops Act 1950 ('any premises from which retail trade or business is carried on') no such definition appears in OSRPA, and it would not therefore be correct to assume that the two Acts should be construed together. Fife and Machin (*Health and Safety at Work* p. 179) take the view that 'shop' is a genus of 'shop premises', and therefore nothing can be a shop unless it also consists of premises. On this view, a mobile shop would be excluded from the Act. The point does not appear to have been authoritatively decided, but, since mobility does not mean immunity (*British Railways Board v Liptrot*, see chapter 4), a shop does not necessarily cease to be one because it is mobile. Further, the word 'shop' has many colloquial uses. Workers in factories frequently refer to their place of work as being a 'shop'; there is a 'closed shop', which is a union membership agreement; one talks about a betting shop, shopfloor, workshop, and even a knocking shop! However, it would appear that the prime characteristic of a 'shop' is the carrying on of retail trade. On this basis, a hairdressing salon which sold goods incidentally to its main business would not be a shop, but a salon in a large departmental store would be.

5.10 (b) A building or part of a building which is not a shop but of which the sole or principal use is the carrying on there of a retail trade or business. The fact that there is a requirement for the business to be 'carried on' implies some degree of permanence, but retail trade means supplying customers for their own use, and would exclude premises used for wholesale purposes (but see (c), below). Retail trade also suggests a dealing in goods, not services. On this interpretation, dry cleaning premises would not be included, though they would be within the meaning of (d) below.

5.11 Retail trade is defined as including off licences, take away food shops, cafes, restaurants, public houses, retail auctions, and private lending libraries (section 1(3)(b)). A building includes a structure (section 90(1)), and thus a permanent market stall (indoor or covered outdoor) would be within the definition, but not a stall erected for temporary purposes, e g a fair or garden fête.

5.12 (c) A building occupied by a wholesale dealer where goods are kept for sale wholesale. Thus a wholesale warehouse or market or display premises fall within the Act.

5.13 (d) A building or part of a building to which members of the public are invited to resort for the purpose of delivering there goods for repair or other treatment, or of themselves carrying out repairs to or treatment of goods. Thus a watch repair shop, a coin-operated laundrette, are within the definition.

5.14 (e) Fuel storage premises which sell solid fuel (coal, coke, etc) but not dock storage or colliery storage premises.

Railway premises (section 1(4))

5.15 This phrase means a building occupied by railway undertakers (i e persons authorised by or under a statute to construct, work or carry on a railway) for the purpose of the railway undertaking carried on by them, but this does not include office or shop premises, premises used for the provision of living accommodation for persons employed in the undertaking, hotels or electrical stations (for which, see Factories Act, section 123).

Canteen or restaurant facilities (section 1(5))

5.16 Any premises which are used in conjunction with office, shop or railway premises for the sale or supply of food or drink to persons employed in the premises shall be treated as being premises of the class in respect of which they are so used. Thus, a canteen which serves office workers shall be treated as being an office, etc.

Cleanliness (section 4)

5.17 All premises and all furniture, furnishings and fittings shall be kept in a clean state. Dirt or refuse must not be allowed to accumulate, and all floors and stairs shall be cleaned not less than once a week by washing or, if it is effective and suitable, sweeping or other method (this does not apply to fuel storage premises which are wholly in the open).

Overcrowding (section 5)

5.18 Rooms shall not be so overcrowded, while work is going on, as to cause a risk of injury to the health of persons working therein. In determining whether or not a room is overcrowded, regard shall be had not only to the number of persons who may be expected to be working in the room at any one time, but also to the space occupied by furniture, fittings, machinery, plant, equipment, appliances and other things.

5.19 The number of persons employed at any time to work in a room shall be such that there is a minimum of 40 square feet of floor space per person, but if the ceiling is lower than 10 feet, then there must be at least 400 cubic feet per person. To make this calculation, it is permissible to ignore the presence of furniture, equipment, etc. However, if the room is one to which the public are invited to resort (e g a shop) the numerical space standards do not apply, but the general requirement not to permit overcrowding remains.

Temperature (section 6)

5.20 Effective provision shall be made for securing and maintaining a reasonable temperature in every room where persons are employed to work otherwise than for a short period, but no method shall be used which results in the escape into the air of any fume of such a character and to such an extent as to be likely to be injurious or offensive to persons working therein. Where the work does not involve physical effort, a temperature of below 16 degrees Centigrade (60.8 degrees Fahrenheit) after the first hour shall not be deemed to be reasonable. However, this does not apply to a room which is office premises where members of the public are invited to resort and in which the maintenance of a reasonable temperature is not reasonably practicable. Nor does it apply to a room which is a shop or railway premises where the maintenance of a reasonable temperature is not reasonably practicable or would cause the deterioration of goods. However, in both these cases, there must be a conveniently accessible and effective means of enabling persons employed to work there to warm themselves, and a reasonable opportunity to use such facilities must be given. For example, in *NAAFI v Portsmouth City Council* the appellants appealed against an Improvement Notice served on them because the temperature in their premises was unreasonably low. It was pointed out that the premises were being used for the storage of fresh food, and a temperature of between 41 and 50 degrees Fahrenheit was necessary. A warm room and suitable warm clothing were provided for employees, and it was held that the Notice should be cancelled.

5.21 On each floor where there is a room to which the Act applies there must be a thermometer, in a conspicuous place and in a position where it can be easily seen and used by persons employed there.

5.22 The Act does not specify a maximum temperature, but this too must be reasonable. Thus if there is an exceptional heatwave, a reasonable temperature must still be observed.

Ventilation (section 7)

5.23 Effective and suitable provision shall be made for securing and maintaining, by the circulation of adequate supplies of fresh air or artificially purified air, the ventilation of every room in which persons are employed.

Lighting (section 8)

5.24 Effective provision shall be made for securing and maintaining sufficient and suitable lighting (whether natural or artificial) in every part of the premises in which persons are working or passing. All glazed windows used for lighting purposes shall, so far as is reasonably practicable, be kept clean on both sides and free from obstruction, except when the windows have been whitewashed or shaded for the purpose of mitigating heat or reducing glare. Artificial lighting shall be properly maintained.

Sanitary conveniences (section 9)

5.25 Suitable and sufficient sanitary conveniences shall be provided at places conveniently accessible to persons employed. These shall be kept clean and properly maintained, with effective provision made for lighting and ventilation. Section 9(5) acknowledges that arrangements may be made for employees to use facilities which are provided for the use of others, and as long as they are conveniently accessible, the Act will have been complied with. In *Davis & Sons v Leeds City Council*, the tenancy of a flat above a shop was subject to the condition that employees at the shop could use the lavatory facilities at any time. An Improvement Notice was issued by an inspector requiring the occupier of the shop to provide accessible sanitary conveniences instead of this arrangement. The industrial tribunal allowed an appeal against the Notice. Evidence was given by a shop assistant that she had not been inconvenienced by this arrangement during her fifteen years' employment, and a circular from the Ministry of Labour addressed to local authorities was admitted into evidence which stated that 'workers in a lock-up shop might have the use of conveniences and facilities in adjacent premises'. After having visited the premises, the industrial tribunal concluded that the lavatory facilities in the flat were conveniently accessible, and cancelled the Notice. However, in *Moorhouse v Effer*, the shared toilet facilities were in a shop which was some distance away and across a busy road, and the industrial tribunal held that these were not conveniently accessible.

5.26 The Schedule to the Sanitary Conveniences Regulations 1964 lays down certain minimum standards as to the numbers of water closets and urinals which must be provided, depending on the number of persons employed, and arrangements are not suitable unless they are at least in accordance with the scales laid down. Public conveniences are not to be taken into account. Lavatories shall not be situated in any room where any person has to work (other than an attendant), and there are further provisions relating to the siting of conveniences, providing effective ventilation, protection from the weather, ensuring privacy, marking for separate sexes, and the disposal of sanitary dressings.

Washing facilities (section 10)

5.27 Suitable and sufficient washing facilities shall be provided at places which are conveniently accessible, including a supply of clean, running hot water and cold or warm water, together with soap and clean towels or other suitable means of cleaning or drying. The place where these facilities are provided shall be kept clean and in an orderly condition, and washing or drying apparatus shall be kept clean and properly maintained. There shall also be effective means for lighting.

5.28 The provisions of this section are complied with if there is suitable shared accommodation, on the lines relating to the provision of sanitary conveniences (above).

5.29 The Washing Facilities Regulations 1964 lay down certain minimum standards as to the numbers of washbasins (or troughs or washing fountains) which must be provided, depending on the number of persons employed. All washing facilities shall be covered and enclosed to ensure sufficient protection from the weather, and effective provision must be made, as far as is reasonably practicable, for ventilation. If there is separate accommodation for each sex, this shall be clearly marked.

5.30 Washbasins, troughs and washing fountains are defined in the Regulations. They must a smooth impervious surface and waste pipes. In *South Surbiton Co-operative Society v Wilcox* an inspector issued an Improvement Notice which required the occupiers to replace a cracked washbasin. This action was confirmed by the industrial tribunal. The surface was not impervious as required by the Regulations, and it was not properly maintained as required by section 10(2). The fact that the company had a previous good record, or that the breach was a trivial one was irrelevant.

Supply of drinking water (section 11)

5.31 An adequate supply of wholesome drinking water must be provided and maintained at suitable places conveniently accessible to persons employed to work in the premises. If it is not piped, it must be contained in suitable vessels and renewed at least daily. All practicable steps must be taken to preserve it and the vessels from contamination. If the water is not supplied from a jet, there shall also be provided (and renewed as often as is required) a supply of drinking vessels of a kind to be discarded after use, or other drinking vessels together with facilities for rinsing them in clean water. Again, it is permissible to use shared facilities which are conveniently accessible.

Accommodation for clothing (section 12)

5.32 At suitable places there shall be suitable and sufficient provision made for enabling persons employed to hang clothing not worn by them during working hours, and such arrangements as are reasonably practicable to enable clothing to be dried. If special clothing is worn at work, and is not taken home, similar arrangements must be made. It would appear that the possibility of there being a theft of clothing is a factor to be taken into account in determining whether or not the accommodation is suitable (*McCarthy v Daily Mirror Newspapers Ltd*).

Sitting facilities (section 13)

5.33 Where persons have, in the course of their work, a reasonable opportunity to sit down without detriment to their work, there shall be provided at suitable places, which are conveniently accessible, suitable facilities sufficient to enable them to take advantage of those opportunities. If persons are employed in shop premises in a room where the public are invited to resort, and have a reasonable opportunity for sitting in the course of their work, the facilities provided shall not be deemed to be sufficient unless there is a ratio of one seat for every three employees. It is an offence if an employer fails to allow his employees to use the seating facilities whenever such use does not interfere with the employee's work.

Types of seats (section 14)

5.34 If the work (or a substantial part of it) can be done sitting down, the seats provided must be of a suitable design, construction and dimension, together with a footrest on which the employee can readily and comfortably support his feet (if he cannot do so without one). The seat and footrest must be adequately and properly supported while in use for the purpose for which it was provided. In *Tesco Stores v Edwards*, the seating provided for assistants at the check-out of a supermarket had no foot rests, were not adjustable, had no back rest, could not swivel, and were generally uncomfortable. An Improvement Notice was confirmed requiring the employers to provide suitable chairs.

Eating facilities (section 15)

5.35 Where persons employed in shop premises eat meals there, suitable and sufficient facilities for eating meals shall be provided.

Floors, passages and stairs (section 16)

5.36 All floors, stairs, steps, passages and gangways shall be of sound construction and properly maintained, and shall so far as is reasonably practicable, be kept free from obstruction and from any substance likely to cause persons to slip. Every staircase must have a substantial handrail which, if there is an open side, must be on that side. In *Haverson & Sons Ltd v Winger* the employer provided a heavy duty rope as a handrail alongside a staircase. An inspector served an Improvement Notice requiring that a rigid handrail be installed in order to comply with section 16, but this was cancelled after an appeal to an industrial tribunal. The Act does not require the handrail to be fixed or rigid, only that it should be substantial, and in the circumstances, a rope handrail was adequate.

5.37 If the staircase has two open sides, or if it is specially liable to cause accidents because of the nature of its construction, or because of the condition of the surface of the steps, or other special circumstances, then the handrail must be provided and maintained on both sides. Any open side of a staircase shall be guarded by the provision and maintenance of efficient means of preventing any person from accidentally falling through the space.

5.38 All openings in the floor shall be securely fenced except in so far as the nature of the work renders such fencing impracticable.

5.39 Section 16 does not apply to fuel storage premises, but in respect of these, the ground surface shall be of good repair, all steps and platforms shall be of sound construction and properly maintained, and all openings in platforms shall be securely fenced except in so far as the nature of the work makes this impracticable.

Fencing (section 17)

5.40 Every dangerous part of any machinery used as, or forming part of, the equipment of the premises shall be securely fenced unless it is in such a position or of such construction as to be as safe to every person working in the premises as it would be if securely fenced. It will be noted that there is no provision in OSRPA corresponding to sections 12 and 13 of the Factories Act relating to prime movers and transmission machinery, and such items must therefore be considered under section 17 of OSRPA. Also, the machinery in question must form part of, or be used as, the equipment of the premises, which is the interpretation placed on section 14 of the Factories Act (*British Railways Board v Liptrot*).

5.41 If a fixed guard cannot be provided, the fencing requirements will be complied with if a device is provided which automatically prevents the operator from coming into contact with the dangerous part. However, no account is to be taken of any person carrying out an examination or lubricating or adjusting any part while in motion which is shown to be immediately necessary, if the examination, lubrication or adjustment can only be carried out while the part is in motion (section 17(3)).

5.42 Fencing shall be of substantial construction, be properly maintained, and kept in position while the parts are in motion or use, except when they are necessarily exposed for examination or lubrication or adjustment (section 17(4)). Thus although the fencing requirements may be discounted when the machine is being examined or lubricated or adjusted, this will only apply when such work is being done by a person over the age of eighteen, and such other conditions as may be specified by Regulations have been complied with. No such Regulations have been made.

Young persons cleaning machines (section 18)

5.43 No person under the age of eighteen shall clean any machinery used as, or forming part of the equipment of the premises if by doing so he is exposed to the risk of injury from a moving part of that machinery or any adjacent machinery. In *Dewhurst Ltd v Coventry Corporation*, a sixteen-year-old boy was cleaning a hand-operated bacon slicer. The cutting edge was partially protected by a guard which had to be removed when the edge was being cleaned. The slicer had no locking device. The boy was instructed to clean one half of the blade, and then rotate it to expose the other half for cleaning. In breach of this instruction he pressed a pad against the cutting edge, and rotated the blade, which resulted in him cutting off the top of his finger. It was held that the employers were guilty of an offence under section 18. Although the blade should not have been moved while cleaning, the cutting edge was still a moving part, and the prohibition in the section was absolute. The employers also failed to establish a defence under section 67 (below) that they had used all due diligence to secure compliance with the section.

Training and supervision (section 19)

5.44 No person shall work on any prescribed machines unless he has been fully instructed as to the dangers arising and the precautions to be observed, and (a) has received sufficient training, and (b) is under adequate supervision by a person who has a thorough knowledge and experience of the machine.

5.45 The Prescribed Dangerous Machines Order 1964 specifies the type of machines in question, which are, generally speaking, mincing, chopping and slicing machines:

5.46 (a) *Machines worked by mechanical power*: worm type mincing machines, rotary knife bowl type chopping machines, dough brakes, dough mixers, food mixing machines when used with attachments for mincing, chipping or any other cutting operation or for crumbling, pie and tart making machines, vegetable slicing machines, wrapping and packing machines, garment presses, opening or teasing machines used for upholstery or bedding work, corner staying machines, loose knife punching machines, wire stitching machines, machines with circular saw blades, band or strip saws, planing machines, verticle spindle moulding machines and routing machines used for cutting wood, wood products, fibre board plastic or similar material.

5.47 (b) *Machines whether or not worked by mechnical power*: circular knife slicing machines used for cutting bacon and other foods (whether similar to bacon or not), potato chipping machines, platen printing machines, including when used for cutting and greasing, and guillotine machines.

Hoists and lifts

5.48 The Offices, Shops and Railway Premises (Hoists and Lifts) Regulations 1968 apply to lifts and liftways which are situated in office premises, shop premises and railway premises to which the Act applies. The obligations under these Regulations fall upon the owner or owners of the building in respect of matters within that person's control or within the control of his servants or agents. Otherwise, the duty to comply with the Regulations falls upon the occupier of the premises.

5.49 Every lift shall be of good mechanical construction, sound material and adequate strength, and properly maintained. They shall be examined be a competent person every six months, and a report containing the prescribed particulars sent to the owner. This report shall be kept for two years and kept readily available for inspection by an inspector. If the examination reveals that the lift cannot continue to be used with safety unless certain repairs are carried out immediately or within a specified time, the person making the report shall also send a copy to the enforcing authority (i e the local authority of HSE).

5.50 Every liftway shall be efficiently protected by a substantial enclosure fitted with gates, and every gate shall be fitted with an interlocking device to ensure that it cannot be opened except when the lift is at a landing, and the lift cannot be moved until the gate is closed.

5.51 Every lift shall be marked conspicuously with the maximum working load, and no greater load shall be carried. Additionally, if the lift is used for carrying persons (as well as goods) efficient automatic devices shall be provided and maintained to prevent the cage overrunning, and each side from which access is reached from the landing shall be fitted with a gate, and an efficient device to ensure that the cage cannot be raised or lowered unless the gate is closed, and will come to rest when the gate is open.

Prohibition of heavy work (section 23)

5.52 No person shall, in the course of his work, be required to lift, carry or move a load so heavy as to be likely to cause injury to him. Whether a load is so heavy as to be likely to cause injury is a question of fact to be decided in each case, depending on the circumstances. As a general guide, it may be said that it is usually not unreasonable for a man to lift a load of up to 145 lbs, and a woman to lift a load of up to a maximum of between 35–50 lbs. In *Hamilton v Western SMT Ltd* two women office workers were required to lift a box which weighed 85 lbs. One woman dropped her end of the box, and the other suffered a back injury. It was held that the employers were liable for damages in respect of the injury.

5.53 A person is not required to lift a heavy load if he lifts it by himself when he has been told to ask for assistance which is readily available (*Peat v N J Muschamp & Co Ltd*). Nor does the fact that an injury has occurred mean inevitably that the employers will be liable. In *Black v Carricks (Caterers) Ltd* the plaintiff was employed as the manageress of a baker's shop. Trays of cakes, bread, etc weighed about 24 lbs, and were stacked in racks. It was the normal practice for two assistants to lift the trays out of the racks so that their contents could be displayed. One morning, none of the assistants arrived for work, and the plaintiff was told to manage as best as she could. While struggling to remove a tray from the rack she slipped and injured her back. Her claim in respect of a breach of statutory duties failed. She had not been 'required' to lift the trays by herself.

5.54 Although it may be considered to be common knowledge that most women cannot lifts weights as heavy as most men, section 23 does not lay down specific weight limits, and it is not therefore a licence to discriminate. In *Cascario v British Railways Board* the employers refused to consider a woman for a job as a railwayman, as this would involve lifting parcels of newspapers which were 40 lb in weight. This was held to amount to sex discrimination, particularly as the woman had been trained in mechanical handling. This case may be contrasted with the later decision of *Page v Freight Hire Tank Haulage Ltd* (see chapter 3).

First aid (section 24)

5.55 This section has been repealed and replaced by the provisions of the Health and Safety (First Aid) Regulations 1981, which will come into force as from July 1982 (see chapter 6).

Multi-occupancy buildings (section 42)

5.56 A building which is under the ownership of one person, but parts of which are leased or licensed to another, must have clean common parts, and all furniture, furnishings and fittings in such common parts shall be kept in a clean state. Effective provision shall be made for securing and maintaining in common parts suitable and sufficient lighting (whether natural or artificial), and all glazed windows and skylights shall, so far as is reasonably practicable, be kept clean on both sides and free from obstruction (except when whitewashed or shaded for the purpose of reducing heat of glare). Floors, stairs, steps, passages and gangways shall be of sound construction and properly maintained, and shall, so far as is reasonably practicable, be kept free from any substance likely to cause any person to slip. Staircases shall be provided with a substantial handrail or handhold, and any open stairway shall be guarded by the provision and maintenance of efficient means of preventing any person from accidentally falling through the space.

5.57 In the event of a contravention of this section, the owner of the building will be guilty of an offence. He will also be liable for a contravention of section 9 and section 10 (see above) where the sanitary conveniences or washing facilities are used in common.

Notification of employment (section 49)

5.58 Before a person begins to employ persons in any office, shop or railway premises, he shall serve on the appropriate authority (i e the enforcement authority) two copies of a notice containing the necessary information. The Schedule to the Notifications of Employment of Persons Order 1964 contains the notice in a prescribed form for use for the purpose of this section. A failure to comply may lead to a fine on summary conviction.

Information for employees

5.59 The Information for Employees Regulations 1965 require the employer to keep posted at all times a copy of the Abstract of the Act and Regulations made thereunder, or to give a copy of the booklet entitled 'Explanatory Book for Employees' (OSR 9A) to persons who work on the premises for more than four weeks. Copies of the Abstract shall be kept in a prominent position, and if posted in the open, kept protected from the weather.

Offences (section 63)

5.60 Since most of the obligations imposed by the Act are placed on the occupier of the premises, it is he who will be primarily liable. If the Act provides that some other person shall be held responsible as well as the occupier, then both shall be liable. The owner is liable in certain multi-occupancy situations (see above).

Defences (section 67)

5.61 It shall be a defence for a person charged with an offence under the Act or a Regulation to prove that he used all due diligence to secure compliance with that provision. This defence is, of course, applicable to criminal proceedings, not to civil liability, although in the latter case a plea of 'no negligence' may equally succeed. 'Due diligence' and 'reasonable care' are similar concepts. It will be recalled that in *J H Dewhurst Ltd v Coventry Corporation* (see para 5.43) the defence of due diligence failed, for although the defendants had given the boy instructions, put up notices, and had received regular visits from the inspectors without adverse comment, the court took the view that in view of the statutory prohibition, there was a special obligation to provide the necessary supervision.

Power to modify agreements (section 73)

5.62 If a person is prevented from doing any structural or other alteration to the premises necessary to comply with a provision of the Act because of the terms of a lease or agreement, he may apply to the local county court, which may make an order modifying or setting aside the terms of that lease or agreement as the court thinks just and equitable. If such alterations do become necessary, the court may also apportion the expense of doing so between the parties having an interest in the premises.

Application of the Act (section 83)

5.63 In so far as the Act imposes duties which may give rise to civil liability (based on an action in tort) the Act is binding on the Crown (except in respect of the Armed Forces). So far as enforcement is concerned, although HSE can inspect Crown premises, it is not possible to prosecute the Crown in the criminal courts.

6 General legal requirements

Employers' liability insurance

6.1 By the Employers' Liability (Compulsory Insurance) Act 1969 every employer (subject to certain exceptions, see below) must take out and maintain an insurance policy with an authorised insurer against liability for bodily injury and disease sustained by his employees, and arising out of and in the course of their employment. The policy must provide cover of at least £2,000,000 arising out of any one occurrence. A contract of insurance is one of *uberrimae fidei*, i e of the utmost good faith, and the proposer must disclose all the information which would affect the mind of a prudent insurer. If, therefore, the employer fails to make such disclosure, the policy may be void, and as well as being without the necessary cover to meet a claim, the employer will be liable for a criminal offence. It must be stressed that the Act applies to every employer carrying on a trade or business, including sports and social clubs, as long as a person is an employee within the legal definition. It is not necessary to take out an insurance policy if the employer engages persons as independent contractors, or domestic servants (who are not employed for the purposes of a business), close relatives, and persons who are not normally resident in Great Britain and who are working here for fewer than fourteen consecutive days.

6.2 The following employers are exempt from the Act:

(a) nationalised industries, local authorities and nuclear power installation operators;
(b) shipowners covered by mutual insurance;
(c) police authorities;

(d) any joint board or committee whose members include representatives of a local authority;
(e) certain other bodies financed out of public funds (see Employers' Liability (Compulsory Insurance) Exemption Regulations 1971).

6.3 A certificate of insurance will be issued by the insurance company, and this must be displayed at each place of business for the information of employees. An inspector from HSE can request to see the certificate, and also the policy itself (on reasonable notice being given). The policy, which must be an approved one (see Employers' Liability (Compulsory Insurance) General Regulations 1971) may be subject to the condition that the insurer may be able to reclaim from the insured any compensation paid to the injured employee from the employer, in specified circumstances, but may not make the employers' liability conditional upon the employer exercising reasonable care, or performing his statutory duty.

6.4 Any employer who is not insured in accordance with the Act shall be guilty of an offence punishable by a fine of £500 for each day when he is in default, and if the offence is committed with the consent, connivance, or facilitated by the neglect of any director, manager, secretary or other officer, then he, as well as the employer (if a body corporate) shall be guilty of an offence and punishable accordingly.

6.5 It is also an offence, punishable by a fine of £200,

(a) to fail to display the certificate or a copy of the insurance certificate;
(b) to fail to send the certificate or a copy to the inspector when so required;
(c) to fail to produce the certificate or a copy on demand by an inspector;
(d) to refuse to allow the inspector to inspect the actual policy document.

6.6 The fact that the employer has a valid policy in force does not confer any automatic right to compensation for an injured employee, as this must be determined by reference to the legal rights of the parties (see chapter 10). However, it does mean that if an employee succeeds in his claim, there will be funds available with which to satisfy the judgment.

6.7 For further information, see *Short Guide to the Employers' Liability (Compulsory Insurance) Act 1969*, available from HSE or Department of Employment.

Fire certificates and fire precautions

6.8 As a guide, it is the intention that general fire precautions shall be the responsibility of the local fire authorities (with Crown premises coming under the jurisdiction of HM Inspector of Fire Services) whereas dangerous materials and processes shall be under the control of HSE. This split of authority was one of the recommendations of the Holroyd Committee on the Fire Service, and is the pattern which has now been adopted. In respect of work premises, there are now four different catagories. First, those premises which obtained fire certificates prior to 1977 under the provisions of the Factories Act and OSRPA. These will remain in force for the time being. Second, those premises which are used as a place of work after January 1977, and which will now require a fire certificate from the fire authority under the Fire Precautions Act 1971. Third, those premises (known as non-certificated premises) which are not required to have fire certificates for a number of reasons. Fourth, 'special premises' where the occupier must apply to HSE for a fire certificate.

1 Pre-1977 certificated premises

6.9 Any fire certificate issued prior to 1 January 1977 under OSRPA or FA will continue in force until it is amended, replaced or revoked. The relevant requirements of those Acts will continue to apply. These are:

(a) All means of escape shall be properly maintained and kept free from obstruction (FA, section 41(1); OSRPA, section 30(1)).
(b) Any proposal to materially increase the number of persons employed in the premises shall be notified to the fire authority (FA, section 41(3); OSRPA, section 30(3)).
(c) While any person is in premises for the purpose of employment or meals, the doors of the premises or of any room in which he is, and any doors which form an exit shall not be locked or fastened as to be incapable of being opened from the inside (FA, section 48(1); OSRPA, section 33(1)).

(d) All doors which open on to a corridor or staircase from any room in which more than ten persons are employed must open outward (unless they are sliding doors) (FA, section 48(3)).

(e) Every hoist or liftway shall be completely enclosed with fire resistant materials (FA, section 48(4)).

(f) Every window, door or other fire exit must be distinctively and conspicuously marked with a notice printed in letters of an appropriate size (FA, section 48(6); OSRPA, section 33(3)).

(g) An audible fire warning system must be provided and maintained (FA, section 48(7); OSRPA, section 34(1)).

(h) Fire warnings shall be tested or examined at least once every three months and whenever an inspector so requires (FA, section 52(1); OSRPA, section 34(2)).

(i) The date of every such test or examination, and particulars of any defect found, together with details of any remedial action, shall be recorded in the general register (FA, section 52(4)).

(j) The contents of any workroom in which persons are employed shall be arranged so that there is a free passageway to a means of escape in the event of fire (FA, section 48(10)).

(k) Employees must be instructed with the means of escape in the case of fire and with the routine to be followed (FA, section 49(1); OSRPA, section 36 (1)).

(l) Appropriate means for fighting fire shall be provided and maintained, and so placed as to be readily available for use (FA, section 51(1); OSRPA, section 38(1)).

2 Premises used after 1 January 1977

6.10 Section 1 of the Fire Precautions Act 1971 empowers the Secretary of State to designate certain premises in respect of which a fire certificate is required. Among those premises which he may so designate are any premises which are used as a place of work (HSWA, section 78(2)). By the Fire Precautions (Factories, Offices, Shops and Railway Premises) Order 1976, factory premises, office premises, shop premises and railway premises in which persons are employed to work were so designated. Consequently, those sections of the Factories Act, OSRPA etc which deal with fire no longer apply, and are replaced by the certification provisions of the Fire Precautions Act.

APPLICATION FOR A FIRE CERTIFICATE (SECTION 5)

6.11 An application for a fire certificate must be made on the prescribed form to the local fire authority. It must specify the particular use or uses of the premises, and must give such information about the premises as may be prescribed. If the premises consist of part of a building, the applicant must give such information as may be prescribed about the rest of the building is so far as it is available to him. The fire authority may then require the applicant (a) to furnish plans of the building, and (b) if the premises consist of part of a building, to furnish them with such plans of the other parts, in so far as it is possible for the applicant to do so. If the applicant fails to furnish the required plans within such time as may be specified or such further time as the fire authority may allow, the application will be deemed to have been withdrawn.

6.12 The fire authority will then carry out an inspection of the building, and will issue a fire certificate provided four conditions are satisfied. These are:

(a) the means of escape in case of fire with which the premises are provided, and
(b) the means with which the relevant building is provided for securing that the means of escape with which the premises are provided can be safety and effectively used at all material times, and
(c) the means of fighting fire (whether in the premises or affecting the means of escape) with which the relevant building is provided for use by persons in the building, and
(d) the means with which the relevant building is provided for giving warnings to persons in the premises,

are such as may be reasonable in the circumstances in connection with that use of the premises. In other words, the premises, or the building, as the case may be, must be reasonably equipped with means of escape, means of fighting fire, and means for giving warnings.

6.13 If, after an inspection, the fire authority is not satisfied that these four conditions have been met, it will serve a notice on the applicant informing him of this fact, and of any steps which would have to be taken (whether by way of alteration to the premises or any other part of the relevant building) and notify him that it will not issue a fire certificate covering that use unless those steps are taken within a specified time.

6.14 An aggrieved person may appeal against the imposition of any such requirement as a precondition to the granting of a certificate, or against the time limit imposed for the taking of those steps (section 9), but if a certificate covering that use has not been issued within the time specified in the notice which has been served by the fire authority, or such further time as may be granted on appeal, the application will be deemed to have been refused.

CONTENTS OF THE FIRE CERTIFICATE (SECTION 6)

6.15 Every fire certificate will specify:

(a) the particular use or uses of the premises;
(b) the means of escape with which the premises are provided;
(c) the means with which the building is provided for securing that the means of escape with which the premises are provided can be safely and effectively used at all times;
(d) the type, number and location of the means for fighting fire with which the relevant building is provided for use by persons in the building;
(e) the type, number and location of the means with which the relevant building is provided for giving persons in the premises warning in the case of fire,

and may do so by reference to a plan.

6.16 The certificate may also impose such requirements as the fire authority consider appropriate in the circumstances:

(a) for securing that the means of escape in case of fire with which the premises are provided are properly maintained and kept free from obstruction;
(b) for securing that the means of escape with which the building is provided are properly maintained;
(c) for securing that persons employed to work in the premises receive appropriate instruction or training in what to do in the event of fire, and that records are kept for that purpose;
(d) for limiting the number of persons who may be in the premises at any one time;
(e) as to other precautions to be observed in the relevant building in relation to the risk to persons in the premises in case of fire.

The above requirements may apply to the whole or any part of the premises or the whole or part of any building.

OFFENCES (SECTION 7)

6.17 If any premises are put to use as a place of work being premises in respect of which a fire certificate is required, and one is not in force, then the occupier of the premises shall be guilty of an offence. However, this does not apply if an application for a fire certificate has been applied for and it has not yet been granted or refused.

6.18 If a fire certificate imposes a requirement, which is contravened (by reason of something being done or not done to any part of the relevant building) by the occupier or some other person as the fire authority consider it to be appropriate in the circumstances (section 6(5)) then the person responsible for the contravention shall be guilty of an offence. This does not apply to any offence committed where the contravention occurred whilst an appeal was pending against the imposition of the requirement in a certificate (section 9(4)), or an appeal was pending against the decision of the fire authority to make a person responsible for the contravention, or the omission from the certificate of a provision which would relieve the occupier from being liable (section 9(5)). Further, a person cannot be convicted of an offence under section 6(5) unless it is proved that his responsibility for contraventions of the requirements in question had been made known to him before the occurrence of the contravention with which he is charged.

6.19 On summary conviction for either of the above offences, a person may be fined up to £1,000; if the conviction is on an indictment, there can be an unlimited fine, or a term of imprisonment of up to two years, or both.

6.20 A copy of the fire certificate shall be kept at the premises (section 6(8)), and a failure to do so renders the occupier liable to a fine of up to £50 on summary conviction.

6.21 If the premises consist of a part of a building and all the parts are under the same ownership, and the premises are held under a lease or licence, for the reference to the occupier shall be substituted the reference to the owner of the building (Fire Precaution Act 1971 (Modifications) Regulations 1976, regulation 3).

CHANGE IN CONDITIONS (SECTION 8)

6.22 Once a fire certificate has been issued and is in force, a number of changes may take place in the conditions which gave rise to its issuance. First, the fire authority may inspect any part of the relevant building at any reasonable time to ascertain whether there has been any change of conditions by reason of which the means of escape, the fire fighting equipment, or the fire warning equipment have become inadequate in relation to any use of the premises covered by the certificate. If the authority are satisfied that there has been such a change, they may serve on the occupier a notice informing him of that fact, and of the steps he would have to take, in relation to the building (whether by way of making alterations or otherwise) to make the matters adequate in relation to that use, and also notify him that if those steps are not taken within such period as is specified with the notice, the fire certificate will be cancelled. Once the steps are taken, the fire authority will issue a new or amended certificate.

6.23 Second, there may be a change in the legal requirements as a result of a Regulation being passed, which would make it appropriate to amend an existing certificate by varying or revoking any requirement which has been imposed under section 6(2), or add to the requirements which the certificate already imposes, or to alter the effect of the certificate as to the person who is responsible under section 6(5) for any contravention. In these circumstances, the fire authority may make such amendments to the certificate to give effect to these changes. However, if the change is as to the responsibility of any person, there must be prior consultation with him if he would be adversely affected by the change.

6.24 Third, there may be a change in the premises consequent on a material structural alteration, a material extension, an internal rearrangement, or an alteration in the furniture or equipment with which the premises are provided, or the keeping of explosive or highly flammable materials anywhere under, in or on the premises in a quantity greater than that already prescribed. In these circumstances the occupier shall, before he carries out such proposal, give notice to the fire authority, and a failure to do so will mean that he is committing an offence.

6.25 A similar provision applies where a fire certificate is in force for a part of a building, and an occupier of another part of that building who is responsible under section 6(5) for the contravention of any requirement imposed by the certificate, should be proposed to keep explosive of highly flammable material in a quantity greater than prescribed. He too, shall notify the fire authority, or he too will be guilty of an offence.

6.26 If, on receipt of the above notices the fire authority are satisfied that the proposed change would render the means of escape, the fire fighting equipment, or the fire warnings inadequate, then they will serve a notice on the occupier within two months, informing him of the steps which must be taken to prevent the matters in question from becoming inadequate, and give such directions as they think fit that the proposed alterations shall not be proceeded with until the steps specified in the notice have been complied with. A failure to so comply will constitute an offence, and the fire authority may cancel the existing fire certificate, whether or not proceedings are brought against the occupier.

APPEALS (SECTION 9)

6.27 A person who is aggrieved by a decision relating to a fire certificate may appeal to the local magistrates' court within twenty-one days from the relevant date. The grounds for appealing are as follows:

(a) against a requirement contained in a notice under section 5(4). This arises when the fire authority have inspected the premises, and, not being satisfied that the statutory requirements have been met, serve a notice informing the occupier what further steps should be taken to satisfy those requirements, or against the time limit imposed by the fire authority for the taking of those steps; or

(b) against the refusal by the fire authority to issue a fire certificate; or

(c) against the inclusion of anything in the fire certificate (or the omission of anything); or

(d) against the refusal of the fire authority to cancel or amend a fire certificate; or

(e) against a direction by the fire authority consequent on any change in conditions; or

(f) against a notification under section 8(5) that unless the occupier takes steps to make adequate the matters specified, the fire certificate may be cancelled, or against the period allowed for the taking of such steps;

(g) against an amendment or cancellation of a fire certificate under section 8(6) brought about by changed conditions or the coming into force of any Regulations.

6.28 The appeal must be made within twenty-one days from the relevant date, which is the date on which the appellant was first served by the fire authority with the notice or refusal, cancellation, direction, amendment or other matter in question, or, in respect of (c) above, the date when the inclusion or omission was made known to him. Once an appeal has been lodged against a refusal to issue a certificate, or against the cancellation or amendment of one already issued, the appellant is not guilty of any offence merely because he used the premises between the relevant date and the date of the final determination of the appeal. If an appeal is brought against the inclusion of a requirement, no offence is committed by reason of a contravention between those dates. When the appeal is brought against the inclusion of a provision making another person responsible under section 6(5) (above), or the omission which, if included, would prevent the occupier being responsible, then again no offence is committed by reason of a contravention between those dates.

6.29 A further appeal will lie to the Crown Court (section 27).

EXCLUSIONS

6.30 The following premises do not require fire certificates, namely, factory premises, office premises, shop premises, and railway premises, in which

(a) not more than twenty persons are employed to work at any one time, or
(b) not more than ten persons are so employed to work elsewhere than on the ground floor.

6.31 However, this exclusion will be lost if

(a) the premises are in a building which contains two or more such premises and the aggregate of those employed to work at any one time in all those premises exceeds twenty; or
(b) the premises are in a building which contains two or more premises and in all those premises the aggregate of persons employed to work at any one time elsewhere than on the ground floor exceeds ten (in other words, in multi-occupancy buildings, it is the total employed in the building which counts, not the separate numbers in each premises); or
(c) in the case of factory premises, explosives or highly flammable materials are stored or used in, under or on the premises. However, the fire authority may decide that the materials are of such a kind or in such small quantity as not to constitute a serious additional risk to persons in the

premises in case of fire. Further exemptions may be made by Regulations under HSWA (see Part II of Schedule 1 of Fire Certificate (Special Premises) Regulations 1976, below).

3 Non-certificated premises (section 12)

6.32 If premises are excluded from the need for a Fire Certificate under the Act on the above grounds, then they are non-certificated premises. Section 12 of the Act enables the Secretary of State to make Regulations about fire precautions which are wide ranging in nature, and under this power, the Fire Precautions (Non-Certificated Factory, Office, Shop and Railway Premises) Regulations 1976 were made. These lay down the precautions which must be observed in those premises in respect of which a fire certificate is not required.

6.33 In factory premises, the following precautions must be observed:

(a) doors opening on to a staircase or corridor from any room in which more than ten persons are employed, and all other doors affording a means of access to or exit from the premises shall open outward (except in the case of sliding doors);
(b) every window, door or other exit affording a means of escape in the case of fire (other than a means of exit in ordinary use) must be distinctively and conspicuously marked by a suitable notice;
(c) every hoist or liftway shall be completely enclosed by a construction which has a fire resistance of not less than thirty minutes, and all means of access to the hoist or liftway shall be fitted with doors having a similar resistance.

6.34 In all factory, office, shop and railway premises, the following precautions must be observed:

(a) the doors through which a worker might have to pass in order to get out of the premises must not be so locked or fastened that they cannot be easily and immediately opened by him on his way out;
(b) the contents of any workroom shall be so arranged as to afford a free passageway to a means of escape in case of fire;
(c) there shall be provided and maintained appropriate means of fighting fire, so placed as to be readily available for use.

Special premises

6.35 The Fire Certificate (Special Premises) Regulations 1976 apply to those premises where there are special hazards due to the storage, manufacture or use of highly flammable liquid, expanded cellular plastics, liquified natural gas, oxygen, chlorine, ammonia, explosives, accelerated charged particles, unsealed radioactive substances, and so premises constructed for temporary occupation for the purpose of building operations or engineering construction.

6.36 An application for a fire certificate in respect of these premises shall be made to HSE, and provided they are satisfied that that there are adequate means of escape, and that the fire fighting equipment and fire warnings are adequate, a certificate will be granted. If they are not so satisfied, they will serve a notice on the applicant, stating what remedial steps must be taken, and indicating that a fire certificate will not be issued unless those steps are taken within a specified time or such further time as may be allowed. There are further provisions relating to the contents of the fire certificate, which may impose conditions, and also dealing with the situation which may arise when changes are made which may affect the matters contained in the certificate, or when, in the opinion of HSE the certificate has become inadequate, or when it needs to be varied.

6.37 An appeal to the magistrates' court may be made by a person aggrieved by anything mentioned in the notice prior to the issue of the certificate, or from a refusal to grant one, or the inclusion or omission of anything in the certificate, etc.

Other fire Regulations

6.38 Existing Regulations which require fire precautions to be observed in industries where there are special hazards still remain in force. Among these, we may note:

(a) Celluloid Regulations 1922, which deal with the manufacture, manipulation, storage and disposal of celluloid and celluloid waste;
(b) Chemical Works Regulations 1922, which lay down certain precautions to be taken when inflammable gases dust and vapours are present;
(c) Manufacture of Cinematograph Film Regulations 1928 which lay down the fire precautions to be observed in rooms in which cinematograph films are made, repaired, manipulated or used;

(d) Cinematograph Film Stripping Regulations 1939, dealing with fire precautions in rooms where cinematograph films are stripped, dried or stored;
(e) Highly Flammable Liquids and Liquified Petroleum Gases Regulations 1972, which lay down the precautions to be taken when premises are used in connection with highly flammable liquid or liquid petroleum gas;
(f) Magnesium (Grinding of Castings and other Articles) Special Regulations 1946, which prohibit smoking and open lights and fires;
(g) Factories (Testing of Aircraft Engines and Accessories) Special Regulations 1952, which lay down precautions to be taken in the event of the leakage or escape of petroleum spirit.

Other legal requirements (section 9A)

6.39 Irrespective of the question of fire certificates, there is a general legal requirement that in respect of all office, shop or railway premises, they shall be provided with such means of escape in case of fire for persons employed to work there as may reasonably be required in the circumstances of the case. In determining this, regard shall be had to the number of persons who may be expected to be working in the premises at any time, and also the number of persons who are not working there, but who may be expected to be there at any time (e g visitors, shoppers, etc).

Power of the court (section 10)

6.40 If the fire authority are satisfied that the risk to persons in the case of fire is so serious that, unless steps have been taken to reduce the risk to a reasonable level, the use of the premises ought to be restricted or prohibited, they may make an application to the court. If the court is similarly so satisfied, it may by order prohibit or restrict the premises from being so used, to the extent appropriate in the case, until such steps have been taken as, in the opinion of the court, are necessary to reduce the risk to a reasonable level.

Working with lead

6.41 The Control of Lead at Work Regulations 1980 came into force in August 1981 and are designed to extend and rationalise the statutory provisions which deal with exposure to lead at the workplace. All existing statutory provisions dealing with lead (section 75 and sections 128–130 of Factories Act) and about twenty Regulations made under the Act have thus been repealed in part or in whole, to be replaced by the new Regulations and an Approved Code of Practice, which sets out acceptable methods of meeting the new requirements. The objective is to protect the health of workers who are exposed to lead dust, fume or vapour, by controlling and reducing the amount of absorption to an acceptable level, and to monitor the amount of lead absorbed by an individual so that he/she may be removed or temporarily suspended from work before health is adversely affected.

6.42 The Code of Practice lays down two standards.

(a) Lead in the air. This remains unchanged and is expressed to be 0.15 milligrams per cubic metre of air expressed as an 8 hour time-weighted average concentration. A higher value is given to tetraethyl lead, which has greater toxicity.

(b) Lead in blood. This is now set at 80 micrograms per 100 millilitres of blood. However, women of child-bearing age have greater vunerability to lead poisoning, and in their case the standard is reduced to 40 micrograms per 100 millilitres of blood.

6.43 If the level of atmospheric lead exceeds half of the above standard, there will have to be regular monitoring; if the blood level content of lead exceeds 40 micrograms blood tests must be taken on a regular basis. Thus once the danger levels have been identified, action can be taken.

LIMITATIONS

6.44 There is some work where the exposure to lead is regarded as being minimal, and unlikely to cause any health problems. These include low temperature melting of lead, such as plumbing, soldering, linotype and monotype casting, painting with low solubility paints, work with material which does not contain more than 1 per cent lead, work with lead in emulsion or paste form so that lead dust or fume cannot be given off, the handling of clean solid metallic lead (ingots, pipes, sheets, etc) and lead from petrol driven vehicles. In these cases, the Regulations will not apply, but reference should be made to the recommendations in the Code of Practice.

6.45 The Regulations impose duties on the employer not only with regard to his employees, but also, so far as is reasonably practicable, in respect of any other person who is at work on the premises where work with lead is being carried on, and who is consequently liable to be exposed to lead from that work.

ASSESSMENT

6.46 Where any work may expose any person to lead, the employer (or self-employed person) shall assess that work to determine the nature and degree of exposure. This must be done within four weeks of the Regulations coming into force (August 1981) in respect of existing work, and in the case of work involving lead commencing after that date, the assessment shall take place before the work commences. The assessment shall be revised whenever there is reason to suppose that the previous assessment is incorrect, or where there is a material change in the work, or when requested to do so by an inspector.

INFORMATION, INSTRUCTION AND TRAINING

6.47 Adequate information, instruction and training must be given by each employer to his employees who are liable to be exposed to lead, so that they are aware of the risks and precautions which must be observed. Similar information, instruction and training must be given to those employees who carry out assessment, cleaning, provide or maintain control measures, or carry out air monitoring (see below).

CONTROL MEASURES

6.48 Every employer shall, so far as is reasonably practicable, provide such control measures for materials, plant, processes as will adequately control the exposure of his employees to lead, otherwise than by the use of respiratory protective equipment or protective clothing by them.

RESPIRATORY EQUIPMENT

6.49 Every employer shall provide each employee who is likely to be exposed to airborne lead with such respiratory protective equipment as is approved by HSE, unless the control methods prove to be adequate.

PROTECTIVE CLOTHING

6.50 Every employer shall provide each employee who is liable to be exposed to lead with adequate protective clothing unless the exposure is not significant.

WASHING AND CHANGING FACILITIES

6.51 Every employer shall provide:

(a) adequate washing facilities;

(b) where protective clothing is provided, adequate changing facilities and facilities for the storage of protective clothing and personal clothing not worn during working hours.

6.52 This requirement does not apply to the first twelve months after the Regulations come into force (August 1981) if it is necessary to erect new buildings or to make substantial alterations to existing buildings in order to comply with the Regulations.

EATING, DRINKING AND SMOKING

6.53 Employers shall take adequate steps to ensure that

(a) so far as is reasonably practicable, his employees do not eat, drink or smoke in any place which is liable to be contaminated by lead;

(b) suitable arrangements are made for employees to eat, drink or smoke in a place which is not liable to be contaminated by lead.

6.54 An employee also commits an offence if he eats, drinks or smokes in any place which he has reason to believe to be contaminated by lead.

CLEANING

6.55 Adequate steps shall be taken to secure the cleanliness of workplaces, premises, plant, respiratory protective equipment and protective clothing.

AVOIDING CONTAMINATION

6.56 Every employer, his employees, and every self-employed person shall, so far as is reasonably practicable, prevent the spread of contamination by lead from the place where work is being carried out.

USE OF CONTROL MEASURES

6.57 Employers who provide control measures, respiratory protective equipment, protective clothing or other thing or facility shall ensure, so far as is reasonably practicable, that they are properly used or applied.

6.58 Every employee shall make full and proper use of any control measure and other thing provided, and if he discovers any defect he shall report it forthwith to his employer.

MAINTENANCE OF CONTROL MEASURES

6.59 All control measures provided by the employer shall, so far as is reasonably practicable, be maintained in an efficient state, in efficient working order, and in good repair.

AIR MONITORING

6.60 Every employer shall have adequate monitoring procedures to measure the concentrations of lead in air to which his employees are exposed, unless the exposure is not significant, and he shall also measure the concentration of lead in air in accordance with those procedures.

MEDICAL SURVEILLANCE AND BIOLOGICAL TESTS

6.61 If an employee is employed on work which exposes him to lead, he must be under medical surveillance by an employment medical adviser or appointed doctor if the exposure to lead is significant, or if the adviser or doctor certifies that he should be under such surveillance. The certificate may state that the employee should not be employed on work which exposes the employee to lead or that he should only be employed subject to certain conditions. Every employee shall present himself for a medical examination or biological tests during working hours when required to do so by his employer.

RECORDS

6.62 Every employer shall ensure that adequate records are kept of the assessment, maintenance, air monitoring, medical surveillance and biological tests, and make these available for inspection by his employees. This does not enable a person to see the health record of an identifiable individual. Entries in these records shall be kept for two years.

Safety signs

6.63 There are a number of legislative provisions which require the display of notices relating to issues of safety and/or health. These include:

(a) Hydrogen Cyanide (Fumigation of Buildings) Regulations 1951 and the Hydrogen Cyanide (Fumigation of Ships) Regulations 1951, which require notices warning of the presence of poisonous gas.
(b) Coal and Other Mines (Managers and Officials) Regulations 1956, which require a danger sign around dangerous areas.
(c) Stratified Ironstone, Shale and Fireclay Mines (Explosives) Regulations 1956 which require warning notices when shotfiring is being carried on.
(d) Quarries (Explosives) Regulations 1959, which require notices warning that shotfiring is about to commence.
(e) Gravel and Sand Quarries (Overhanging) (Exemption) Regulations 1958, which require warning notices where there is danger from falls.
(f) Ionising Radiations (Unsealed Radioactive Substances) Regulations 1968, and the Ionising Radiations (Sealed Sources) Regulations 1969 require warning notices around the boundaries of radiation areas.
(g) Chemical Works Regulations 1922 require notices prohibiting smoking and the use of naked lights where there is a likelihood of explosion.
(h) Factories (Testing of Aircraft Engines) Special Regulations 1952 require notices prohibiting smoking in test areas.
(i) Shipbuilding and Ship Repairing Regulations 1960 require warning notices when persons are working or passing underneath any place from which articles may fall, and notices prohibiting smoking or the use of naked lights near acetylene generating plants.
(j) Highly Flammable Liquids and Liquified Petroleum Gases Regulations 1972 require various notices warning persons of the presence of flammable liquids, and prohibiting smoking near such liquids.
(k) Health and Safety (Agriculture) (Poisonous Substances) Regulations 1975 require warning notices outside greenhouses stating that a specified substance is being used.

Safety Signs Regulations 1981

6.64 These Regulations came into force on 1 January 1981, following an EC Directive designed to standardise safety signs throughout the Community. They are only in force in respect of new signs; because of the high cost of changing existing signs a five year period has been allowed for during which all existing signs may be progressively changed to the new format. The Regulations do not require signs to be erected; they only require that if signs are erected, they must be in the new form. The new symbols will apply to all places of employment except for coal mines, although internal works road signs (including those relating to pedestrians) must conform to Road Traffic Act signs. Signs which are placed near roads or railways must not be put where they can be confused with signs for the control of the traffic.

6.65 The new Regulations do not cover certain matters for which there are as yet no internationally agreed symbol, e g emergency exits, fire-fighting equipment etc, and any existing legal requirements should continue to be observed.

6.66 There are four catagories of signs, each with its own distinctive shape and colour.

i *Prohibition signs*
6.67 These will be circular with a red border and crossbar over a black symbol on a white background. It is meant to indicate a hazard which must not be ignored, e g 'No Smoking'.

ii *Warning signs*
6.68 These will be triangular in shape with a black border and symbol on a yellow background. It will denote a hazard which could be fatal if overlooked, e g 'Caution'.

iii *Mandatory signs*
6.69 These will be circular on a blue background with symbols in white. These will indicate specific instructions that must be obeyed, where there is an obligation to use safety equipment, e g 'Hearing protection must be worn'.

iv *Safe conditions*
6.70 These will be square or oblong (depending on the size of the text or symbol) and will consist of a green background with white symbols. These will denote some safety consideration, e g 'First Aid post'.

Examples of safety signs and their meanings

Prohibition signs

Do not extinguish with water

No smoking

Warning signs

Caution, toxic hazard

Caution, industrial trucks

Mandatory signs

Hearing protection must be worn

Foot protection must be worn

Safe condition signs

First aid

Indication of direction

Labelling of articles and substances

6.71 There are a number of statutory provisions which require the marking or labelling of articles or substances, or the boxes, containers, receptacles etc which contain them. Among these are:

(a) Asbestos Regulations 1969;
(b) Carcinogenic Substances Regulations 1967;
(c) Celluloid and Cinematograph Film Act 1922;
(d) Celluloid Regulations 1921;
(e) Cinematograph Film Stripping Regulations 1939;
(f) Manufacture of Cinematograph Film Regulations 1928;
(g) Lead Paint Regulations 1927;
(h) Petroleum (Consolidation) Act 1928;
(i) Petroleum (Carbide of Calcium) Order 1929;
(j) Petroleum (Mixtures) Order 1929;
(k) Petroleum Spirit (Motor Vehicles) Regulations 1929;
(l) Petroleum (Liquid Methane) Order 1957;
(m) Petroleum (Carbon Disulphide) Order 1968.

Dangerous substances

6.72 Under the provisions of the Packaging and Labelling of Dangerous Substances Regulations 1978 (as amended) suppliers of certain specified industrial chemicals must put warning labels on containers to enable the user to speedily identify the main dangers, and take the necessary precautions in handling and using. The chemicals must be in containers which are properly designed, constructed and secured in order to prevent accidental spillage in normal use.

6.73 The warning labels will consist of a pictorial symbol and a key word. The symbol will be printed in black on an orange or yellow background so that it stands out, and will thus identify the likely danger. The Schedule to the Regulations lists over 800 chemicals and a further 121 were added by the Packaging and Labelling of Dangerous Substances (Amendment) Regulations passed in 1981. The chemicals are classified according to the main hazard, e g explosives, highly flammable, corrosive, toxic, irritant, harmful, oxidising and any combination of these hazards. Each label will carry risk phrases which spell out in detail the main dangers, and will give advice on sensible safety precautions. The name of the chemical will be stated, and also the name and address of the supplier or manufacturer who may assist in the event of an accident.

These Regulations are designed to ensure compliance with the EC Directives on this subject (see chapter 11).

The reporting and notification of accidents and dangerous occurrences

6.74 The Notification of Accidents and Dangerous Occurrences Regulations 1980 came into force on 1 January 1981, and lay down a new procedure for reporting and notification. Consequently, certain legislation on this subject has now been repealed, i e section 80, Factories Act; section 48, OSRPA; sections 116–120, MQA. The result is that all occupational accidents and dangerous occurrences come within the new Regulations. These apply to all work activities covered by HSWA in Great Britain, and to certain specialised work in territorial waters, but they do not apply to offshore installations or to pipeline work within territorial waters.

6.75 The Regulations deal with four types of incidents, namely fatal accidents, major injuries, dangerous occurrences, and other accidents. Only the first three of these are reportable to the enforcing authority, and the last one will be notified through the DHSS industrial injuries claims procedures.

A FATAL ACCIDENTS

6.76 A fatal accident must be reported when it arises out of or in connection with any work activity, and irrespective of whether or not the deceased was employed or not. Thus, if a member of the public is killed as a result of activities at work, this must be reported. Examples would be if there is a collapse of scaffolding on a public highway, or an incident in a factory, which causes the death of an employee or a member of the public. Further, if there has been a notifiable accident or dangerous occurrence which has been reported, and a death of an employee occurs within a year thereof, it must be reported, whether or not it has already been reported. This provision may cause some difficulty in practice, as the person in question may have ceased to be an employee at the time of the death, or the person upon whom the duty to report falls may be unaware that the death has taken place. All he can do in such circumstances is to take reasonable steps to keep himself informed of the progress of the injured person, and while this would not excuse a failure to report, it may avoid a prosecution, or mitigate any punishment.

B MAJOR INJURY

6.77 This is defined as being

(a) fracture of the skull, spine or pelvis;
(b) fracture of any bone
  (i) in the arm, other than a bone in the wrist or hand,
  (ii) in the leg, other than a bone in the ankle or foot;

(c) the amputation of any foot or hand;
(d) the loss of sight in an eye;
(e) any other injury which results in the person injured being admitted to hospital as an in-patient for more than twenty-four hours, other than purely for observation.

6.78 Again, such accidents must be reported, whether they are suffered by an employee or a member of the public. The extent of the injury is irrelevant, and may not even be immediately apparent. Further, the injured person may not be taken to hospital immediately. Some follow up procedure, therefore should be established to ascertain the extent and likely development of a particular injury.

C THREE DAY ACCIDENTS

6.79 If an accident (other than death or major injury) to an employee while at work results in his being incapacitated for more than three consecutive days (excluding the day of the accident and a rest day, e g Sunday) then such accident will no longer be reportable. Instead, the information will come to HSE from the DHSS if the employee makes a claim for industrial injuries benefit under section 88 of the Social Security Act 1975 (see chapter 9). When a claim is made, the employer will have to submit a report on Form BI 76 (or BI 77 for industrial diseases), and a copy will be sent to HSE.

D DANGEROUS OCCURRENCES

6.80 Certain dangerous occurrences must be reported whether or not any person is killed or injured. The complete list can be found in Schedule 1 to the Regulations, which is divided into four parts.

i *All dangerous occurrences*

6.81 Part I deals with dangerous occurrences which are reportable wherever they occur. These include:

(a) the collapse or overturning of any lift, hoist, crane, excavator, or mobile powered platform, the failure of any load bearing part of a platform which might have cause a major injury to any person;
(b) any explosion, collapse or bursting of any closed vessel including a boiler or boiler tube in which there was any gas (including air) or vapour at a pressure greater than atmospheric, which might have caused major injury to any person, or which did result in significant damage to plant;
(c) electrical short circuit or overload accompanied by fire or explosion which resulted in the stoppage of the plant for more than twenty-four hours, and which might have caused a major injury to any person;

(d) an explosion or fire occurring which resulted in the stoppage of plant or suspension of normal work for more than twenty-four hours, where such explosion or fire was due to the ignition of process materials, their by-products (including waste) or finished products;

(e) the sudden uncontrolled release of one tonne or more of highly flammable liquid, flammable gas, or flammable liquid above its boiling point;

(f) a collapse of any scaffolding which is more than 12 meters high which results in a substantial part of the scaffolding falling or overturning;

(g) the collapse of any building or structure under construction, reconstruction, alteration, or demolition, involving a fall of more than 10 tonnes of material, except when this was intentional;

(h) the uncontrolled release of any substance which might be liable to cause damage to the health of, or major injury to, any person;

(i) any incident where a person is affected by the inhalation, ingestion or absorption of any substance, or by lack of oxygen, likely to cause ill-health requiring medical treatment;

(j) any case of acute ill-health resulting from occupational exposure to isolated pathogens or infested material;

(k) the unintentional explosion of explosives;

(l) failure of any freight container while being raised, lowered or suspended;

(m) the bursting, explosion, or collapse of a pipe-line or the ignition of anything in the pipe-line;

(n) the overturning of any road tanker, or where a road tanker suffers serious damage while conveying prescribed hazardous substances.

ii *Dangerous occurrences in mines*

6.82 Part II of the Schedule deals with dangerous occurrences in mines. These include:

(a) the ignition of any gas or dust;

(b) the accidental ignition of any gas in part of a fire damp-drainage system;

(c) the outbreak of any fire below ground;

(d) any incident whereby because of smoke a fire may have broken out causing a person to leave any place below ground;

(e) the outbreak of any fire on the surface endangering the operation of certain equipment;

(f) any violent outburst of gas;
(g) the breakage of any rope, chain, coupling or other gear by which persons are carried;
(h) the breakage of any rope, chain, coupling or other gear used for the transport of persons below ground or the breakage of any man-carrying conveyor belt;
(i) incidents which relate to the overwinding of any cage;
(j) the breakdown of ventilation apparatus;
(k) the collapse of any head-frame, winding engine house, screen or tippler house or vehicle gantry;
(l) an incident where breathing apparatus fails to function while being used, or any person receives first aid or medical treatment after using such apparatus;
(m) an incident in which a person suffers electric shock or burns requiring medical treatment from any voltage exceeding 25 volts;
(n) an injury resulting from an explosion or discharge of blasting material for which he receives first aid or medical treatment;
(o) an incident where emergency apparatus is used (other than in training or practice);
(p) an inrush of noxious or flammable gas from old workings;
(q) any inrush of water from any source;
(r) any movement or other event which indicates that a tip is likely to become insecure.

iii *Dangerous occurrences in quarries*

6.83 Part III of the Schedule deals with occurrences which are notifiable in relation to quarries. These are:

(a) the collapse of any load-bearing structure;
(b) the sinking or overturning of any waterborne craft;
(c) an explosion which results in an injury requiring first aid or medical treatment;
(d) blasting operations which result in something being projected beyond the quarry which may endanger persons;
(e) an incident in which a person suffers electrical shock requiring first aid or medical treatment where the voltage exceeds 25 volts;
(f) any movement or other event which indicates that a tip is likely to be insecure.

iv *Dangerous occurrences on railways*

6.84 Part IV of the Schedule deals with dangerous occurrences which are notifiable in relation to railways. These are:

(a) certain incidents which endangered or were likely to endanger the safety of the passengers or crew of a train;

(b) collisions, derailment or dividing of a train (except during shunting operations);
(c) failure of equipment at a level crossing.

Repeals

6.85 Existing requirements to report dangerous occurrences have been revoked. For example, it is no longer necessary to report the bursting of an abrasive wheel under power in a factory. The Dangerous Occurrences (Notification) Regulations 1947, and a number of other provisions have been revoked in whole or in part (see Schedule 5).

The reporting of accidents and dangerous occurrences

6.86 The responsible person must report the accident or dangerous occurrence. Usually, this will be the employer, although in the case of mines, quarries, closed tips or pipelines, this will be the mine manager or owner, as the case may be. If a member of the public is injured, the responsible person will be the occupier of the premises at which the incident took place.

6.87 The reporting must be to the appropriate authority by the quickest practicable means. As a general rule, this would mean the use of the telephone. Within seven days, a report must be sent in on Form F2508 to the enforcing authority, giving as much information as possible. This will enable a prompt investigation to begin.

6.88 The appropriate authority will be the following:

Premises by main activity	*To whom to report*
(1) Factories and factory offices	Factory inspector
(2) Mines and quarries	Mines and Quarries inspector
(3) Farms, horticultural premises and forestries	Agriculture inspector
(4) Civil engineering and construction	Factory inspector
(5) Statutory and non-statutory railways	Inspecting officer of railways
(6) Offices, shops, separate catering services, laundrettes	Local authority
(7) All other premises (hospitals, research and development establishments, water, postal services, entertainment and recreational services, local government premises, education and road services	Factory inspector

Exemptions

6.89 (a) Certain existing reporting requirements have been retained, and consequently there is no need to report under the new Regulations. These include accidents on railways, nuclear installations, and aircraft. Explosions caused by agricultural substances are also excluded. For a full list, see Schedule 2.

6.90 (b) It is not necessary to report any accident to a self-employed person who is not engaged in work under the control of another person.

6.91 (c) It is not necessary to report any accident to a person which occurred while he was a patient undergoing treatment in a hospital or the surgery of a doctor or dentist, to a member of the Armed Forces or visiting forces who was on duty at the time of the accident.

6.92 (d) It is not necessary to report road accidents, unless a person is killed while engaged on work alongside or on the road, or while working on the construction, demolition, alteration, repair or maintenance of the road or markings on the road or equipment, or the verges, fences, hedges, or boundaries of the road, or working on the pipes or cables which are on, under or over the road, or whilst working on buildings or structures adjacent to the road. In other words, normal road accidents will be reported to and investigated by the police in the usual way. It will only be necessary for HSE to become involved if the accident was a work accident which occurred on the roadway generally.

Records

6.93 Records of all notifiable accidents and dangerous occurrences must be kept by every employer. These shall be kept at the place where the work to which they relate is carried on, or, if that is not reasonably practicable, at the usual place of business. Form 2509 may be used for this purpose, but an employer may use his own register, provided it contains all the information required by Schedule III. This information includes:

(a) the date of the accident or dangerous occurrence;
(b) particulars of the person injured, i e name, sex, age, occupation, and the nature of the injury;
(c) place where the accident or dangerous occurrence happened;
(d) a brief description of the circumstances.

6.94 Additionally, the employer must keep records concerning all enquiries from the DHSS which relate to claims in respect of prescribed industrial diseases, or claims concerning pneumoconiosis or byssinosis. The required information is the name, sex, occupation of the person concerned, the nature of the disease in respect of which the claim was made, and the date of the first absence from work in respect of that disease. When an employee makes a claim for industrial injuries benefit, the DHSS will send Form BI 76 to the employer. This must be completed by him and returned, and a copy will then be sent to HSE by the DHSS. To facilitate this arrangement, the new Form BI 76 has been produced in such a way so that by filling in the top copy with a ball-point pen, the under copy will be completed also. In respect of claims for prescribed industrial diseases, Form BI 77 will be used for DHSS purposes. However, in preparing the report for the employer's own records, Part II of Form 2509 can be used.

Safety precautions

6.95 The employer's duty to provide adequate safety precautions stems from three main sources. First, there are a number of legislative provisions which require the employer to provide safety precautions and/or equipment. It will be recalled that by virtue of section 9 of HSWA it is an offence to make a charge for such items. Second, there is the general duty under section 2 of HSWA to ensure, so far as is reasonably practicable, the health, safety and welfare of all employees, and it may be argued that a failure to provide appropriate precautions could amount to an offence under this section. Clearly, the cost, inconvenience and efficacy of the precautions can be taken into account, and balanced against the risks involved (see *Associated Dairies v Hartley*, chapter 3). Third, there is the employer's duty at common law to take care in order to prevent or eliminate the risk of an injury occurring, and a failure to provide the necessary precautions may mean that the employer has failed to fulfil that duty (see chapter 9).

6.96 If an employer does provide the safety precautions, they must be suitable and effective, and conform to acceptable standards. For this purpose, British Standards should be observed where possible. It is further necessary to ensure that the precautions are suitable for each employee to use or wear, for the legal duty is owed to them as individuals, not collectively (*Paris v Stepney Borough Council,* see chapter 9).

6.97 The following list of the more common items should be considered as appropriate:

Head protection: safety helmets, caps, hairnets, faceshields;
Face protection: safety spectacles, goggles, shields;
Ear protection: ear plugs, ear muffs, ear valves;

Breathing protection: respirators, breathing apparatus, masks;
Body protection: overalls, aprons, spats, gloves;
Foot protection: safety footwear, wellingtons, rubber boots.

6.98 It is important to provide appropriate protective clothing in appropriate conditions. This means that 'over-kill' should be avoided. For example, if there is a noise hazard, the protection must match the volume of noise which is hazardous, but not eliminate harmless (or even useful) noise, for this could result in a further hazard being created. An employee who wears every single item of protective clothing which may be available will surely resemble someone from outer space, and be a positive menace to himself and possibly others! Safety audits, sampling and surveys are useful sources of continuing information concerning the efficacy of the precautions.

6.99 The following is a list of various statutory requirements relating to protective clothing and equipment which are currently in force.

Statutory provision	*Requirements*
(1) Aerated Water Regulations 1921	suitable face-guard and gauntlets, waterproof aprons with bibs, waterproof boots or clogs
(2) Agriculture (Poisonous Substances) Act 1952	see Health and Safety (Agriculture) (Poisonous Substances) Regulations 1975, below
(3) Asbestos Regulations 1969	approved respiratory protective equipment and protective clothing
(4) Blasting (Castings and other articles) Special Regulations 1949	approved protective helmets, suitable gauntlets and overalls
(5) Cement Works Welfare Order 1930	water tight thigh boots, suitable goggles, waterproof coats, overalls and headcoverings (for females)
(6) Chemical Works Regulations 1922	approved breathing apparatus, life belt, suitable overalls, protective footwear, protective coverings, respirators non-metallic shoes, gloves
(7) Chromium Plating Regulations 1931	aprons with bibs, loose fitting rubber gloves, rubber boots
(8) Clay Works (Welfare) Special Regulations 1948	suitable protective clothing, oilproof aprons

Statutory provision	*Requirements*
(9) Construction (General Provisions) Regulations 1961	suitable respirators
(10) Construction (Health and Welfare) Regulations 1966	adequate and suitable protective clothing
(11) Control of Lead at Work Regulations 1980	Respiratory equipment, adequate protective clothing
(12) Diving Operations Special Regulations 1960	suitable plant or equipment
(13) Dyeing Welfare Order 1918	suitable protective clothing, loose fitting rubber gloves
(14) East India Wool Regulations 1908	suitable overalls and respirators
(15) Electricity Regulations 1908 (as amended)	insulating boots and gloves
(16) Flax and Tow Spinning and Weaving Regulations 1906	suitable respirators
(17) Foundries (Protective Footwear and Gaiters) Regulations 1971	suitable footwear and gaiters
(18) Fruit Preserving Welfare Order 1919	suitable protective clothing
(19) Glass Bevelling Welfare Order 1921	suitable protective clothing
(20) Grinding of Metals (Miscellaneous Industries) Regulations 1925	suitable respiratory protective equipment
(21) Gut Scraping, Tripe Dressing etc Welfare Order 1920	suitable overalls, waterproof aprons and boots
(22) Health and Safety (Agriculture) Poisonous Substances) Regulations 1975	Rubber gloves, rubber boots, face shield, overalls, rubber apron, mackintosh, eye shield, gauntlet gloves, respirator, hood, dust mask, rubber coat, sou'wester, rubber apron
(23) Hemp Spinning and Weaving Regulations 1907	suitable respirators
(24) Hollow-ware and Galvanising Welfare Order 1921	suitable protective clothing, finger stalls, rubber gloves, aprons, clogs
(25) Horse-hair Regulations 1907	suitable overalls and headcoverings, suitable respirators

Statutory provision	*Requirements*
(26) Ionising Radiations (Unsealed Radioactive Substances) Regulations 1968	personal protective equipment (clothing and breathing apparatus)
(27) Iron and Steel Foundries Regulations 1953	suitable gloves, approved respirators
(28) Jute (Safety, Health and Welfare) Regulations 1948	suitable respirators, overalls, headcoverings
(29) Laundries Welfare Order 1920	suitable protective clothing, overalls or aprons with bibs, armlets
(30) Magnesium (Grinding of Castings) Special Regulations 1946	suitable fire resisting overalls, overalls, leather aprons with leather bibs
(31) Non-Ferrous Metals (Melting and Founding) Regulations 1962	suitable gloves, respirators
(32) Oilcake Welfare Order 1929	suitable protective clothing
(33) Patent Fuel Manufacture (Health and Welfare) Regulations 1946	barrier cream, suitable goggles
(34) Pottery (Health and Welfare) Special Regulations 1950	protective clothing of suitable design, waterproof apron, approved respirators
(35) Protection of Eye Regulations 1974	eye protectors
(36) Sacks (Cleaning and Repairing) Welfare Order 1927	suitable protective clothing
(37) Shipbuilding and Ship Repairing Regulations 1960	breathing apparatus, lamp or torch, hand protection, gauntlets
(38) Tanning (Two Bath Process) Welfare Order 1918	rubber or leather aprons with bibs, rubber boots or leather leggings, rubber gloves
(39) Tanning Welfare Order 1930	protective clothing, aprons, leg coverings, rubber or leather gloves
(40) Tin or Terne Plates Manufacture Welfare Order 1917	suitable waterproof aprons, clogs
(41) Wool, Goat Hair and Camel Hair Regulations 1905	suitable overalls and respirators

6.100 Additionally, recommendations relating to the use of protective clothing or equipment may be found in relevant Codes of Practice, e g 'Vinyl Chloride Monomer Health Precautions', 'Protection of Persons exposed to Ionising Radiations in Research and Teaching' and so on. A recent discussion document issued by HSE suggested that a voluntary Code of Practice should be drawn up for the construction industry, which would recommend, among other things, that the wearing of safety helmets should be made a condition of employment.

6.101 The legal problems involved in enforcing the use of safety clothing and equipment will be discussed in chapter 10.

First aid

6.102 The Health and Safety (First-Aid) Regulations were made in June 1981 and will come into force as from 1 July 1982. Four statutory provisions are thus repealed (Factories Act, section 61; Mines and Quarries Act, section 115 (in part); OSRPA, section 24; and Agriculture (Safety, Health and Welfare Provisions) Act, section 6(1) and (4)). Additionally, some forty-two Regulations and Orders will be revoked, and thus the new Regulations, supported by an Approved Code of Practice and Guidance Notes, will be the main source of legal rules.

6.103 The Regulations lay down two separate general requirements. First, an employer shall provide (or ensure that there is provided) such equipment and facilities as are adequate and appropriate in the circumstances for enabling first aid to be rendered to his employees who are injured or become ill at work. Guidance on the phrase 'equipment and facilities', and when these are 'adequate and appropriate' can be found in the Approved Code of Practice.

6.104 Second, an employer shall provide (or ensure that there is provided) such number of suitable persons as is adequate and appropriate in the circumstances for rendering first aid to his employees if they are injured or become ill at work. A person shall not be regarded as being suitable for this purpose unless he has undergone.

(a) such training and has such qualifications as HSE may approve for the time being in respect of that case, and
(b) such additional training, if any, as may be appropriate in the circumstances of that case.

6.105 However, where such person is absent in temporary and exceptional circumstances, it is sufficient compliance if another person is appointed to take charge of first aid during the period of absence. Further, where, having regard to the nature of the undertaking, the number of employees at work, and the location of the establishment, it would be adequate and appropriate to appoint someone to be responsible for first aid (who may not necessarily be trained or qualified as above) the employer meets the legal obligations by making such appointment.

6.106 The employer must inform his employees of the arrangements that have been made in connection with the provision of first aid, including the location of equipment, facilities, and personnel.

6.107 Self-employed persons shall provide (or ensure that there is provided) such equipment, if any, as is adequate and appropriate in the circumstances to enable him to render first aid to himself while at work.

Approved Code of Practice

6.108 This Code deals with the need to provide the appropriate numbers and types of personnel to render first aid (or to ensure that it is rendered), and the equipment and facilities which will be needed. It then lays down the criteria which should be adopted to determine what each employer should do in the particular circumstances.

PERSONNEL

6.109 The Regulations require that 'suitable persons' shall be appointed. These may be:

(a) *First-aider* This is a person who has been trained by an organisation approved by HSE, and who holds a current first-aid certificate.

(b) *Occupational first-aider* This is a person who has been trained by an organisation approved by HSE, and who holds a current occupational first-aid certificate.

(c) *Other persons* This catagory will include any other person who has undergone training and obtained qualifications approved by HSE, and who may be counted as a first-aider or occupational first-aider.

6.110 The Regulations also require that an appointed person shall be provided by the employer for the purpose of taking charge of the situation (e g to call an ambulance) if a serious injury or major illness occurs, and to be responsible for first-aid equipment, if no first-aider or occupational first-aider has been appointed, or if one is appointed but is absent in exceptional or temporary circumstances. An appointed person is not an acceptable alternative to a first-aider or occupational first-aider, except in those cases where because of the nature of the undertaking (with a low hazard rating) or small number of employees, or location of the establishment, trained first-aid personnel are not required.

EQUIPMENT AND FACILITIES

6.111 The following should be provided as appropriate:

i *First-aid boxes and travelling first-aid kits*

6.112 These should contain a sufficient quantity of suitable first-aid materials and nothing else. The contents should be replenished as soon as possible after use, and items which deteriorate over a period of time should be replaced. Each self-contained working area should have its own first-aid box, which should be clearly identified and be in an accessible location. First-aid kits should be checked frequently to make sure that they are fully equipped and that all items are usable. Guidance as to the contents of boxes and kits is given in the Guidance Notes which accompany the Approved Code of Practice.

ii *First-aid room*

6.113 This would be needed in all establishments where more than 400 employees are at work, or where there are unusual or special hazards, or whenever the employer considers it may be needed because there is dispersed working or access to places of treatment is difficult. A suitably qualified person should be responsible for the first-aid room and its contents (e g an occupational first-aider or a first-aider in some circumstances) and the room should not be used for any other purpose. It should contain suitable facilities and equipment (the Guidance Notes provides advice on these), be clearly identified as such, have suitable waiting facilities provided, and there should be effective means of communication between all work areas, the first-aid room and the first-aider or occupational first-aider on call.

iii *Other equipment*

6.114 Where first-aiders or occupational first-aiders are employed, appropriate equipment should be provided, such as stretchers, carrying chairs, wheel chairs, etc, to enable a sick or injured employee to be removed to a safer or more hygienic environment.

WHAT SHOULD BE PROVIDED IN PRACTICE

6.115 In determining what equipment and facilities an employer should provide, and what numbers and types of suitable persons should be appointed, regard should be had to the following matters:

i *The number of employees exposed to the risk*

6.116 When considering the total number of first-aiders needed, account should be taken of the number of employees at work, the nature of the work undertaken at the establishment, whether the work is undertaken in scattered locations, whether or not there is shift work, and the distance from outside medical services. In establishments with a relatively low hazard, e g offices, shops, banks or libraries, etc, it would not be necessary to appoint a first-aider unless 150 or more employees were at work, and a ratio of one first-aider for each 150 employees would probably be adequate. If there are fewer than 150 employees at work, a first-aider may not be needed, but there must always be an appointed person to take charge at all times when there are employees at work. If there is a greater degree of hazard, such as factories, dockyards, warehouses and farms, a first-aider should be appointed if there are more than 50 persons employed, and if the number of employees is significantly larger, there should be one first-aider for each 150 persons employed. If there are fewer than 50 persons employed, an appointed person should be in charge at all times when persons are at work.

6.117 If the establishment has unusual or special hazards, the employer will need to provide more than the numbers of first-aiders suggested above, and an occupational first-aider should be appointed (who will count as a first-aider for the purpose of determining whether or not an adequate number of first-aiders have been appointed). If there is shift working, sufficient first-aiders should be appointed to provide adequate coverage for each shift, taking into account the number of employees at work on the shift.

6.118 If there are more than 400 employees at work, a suitably equipped and staffed first-aid room should be provided, but if there is an unusual or special hazard (e g shipbuilding, chemical industries, quarries), or the place of work is a long distance from medical services, or transport facilities are limited, consideration should be given to the provision of a first-aid room even though there are fewer than 400 employees at work.

ii *The nature of the undertaking*

6.119 Some processes will involve special hazards, and consequently different first-aid facilities will be needed. If there is a known danger or a recognised potentially harmful substance, the first-aider will need to be specially briefed, and know how to take effective action immediately. If there is a special or unusual hazard, a first-aid room will be needed, and occupational first-aiders given special instructions.

iii *The size of the establishment and the distribution of employees*

6.120 Every employee should have a reasonably rapid access to first aid. In a compact establishment, this might involve a central point, but in a large plant with staff dispersed over a wide area, first-aid equipment and personnel should be located in different parts. This may mean an increase in the numbers of suitable persons than the figures suggested above. If there is a large number of employees in self-contained working areas, the employer may need to provide centralised facilities (e g a first-aid room) and supplementary equipment and personnel in other locations.

iv *The location of the establishment and locations of work*

6.121 If access to emergency medical facilities (e g a hospital) is difficult because of distance or limited transport facilities, consideration should be given to the provision of a first-aid room even when the numbers of employees or nature of the hazards are such that this would not otherwise be justified. Additionally, one or more first-aiders may be provided even though there is a smaller number of employees working there than the numbers suggested above. If employees are working away from the employer's establishment, the employer must still ensure that adequate first-aid provision is made. Obviously, this would depend on all the circumstances, e g if they are working alone or in small groups, the nature of the hazards they are likely to encounter, and so on. For example, if employees work alone or in small groups in isolated locations, or a long way from medical services, or are using potentially dangerous tools or machinery, small travelling first-aid kits should be provided. If there is a large number of employees working in these circumstances, a first-aid box or even first-aid facilities and personnel should be considered, taking into account all the relevant factors. If employees are working with employees of another employer, the two employers may agree between themselves as to the provision of the necessary first-aid equipment, facilities and personnel (e g on construction sites). The agreement should be in writing, and each employer should keep a copy. The employer who is not making the necessary arrangements should ensure that the arrangements made by the other employer are adequate.

7 Particular health and safety problems

Pregnant women

7.1 There are a number of legal restrictions on the hours of work which women may work, and further restrictions on the nature of their employment. These matters are currently under review by the Equal Opportunities Commission and HSC, and are outside the scope of this book. However, there are certain safety considerations which must apply peculiarly to pregnant women because of potential dangers to the foetus. Thus under the Ionising Radiation (Unsealed Radioactive Substances) Regulations 1968 (see para 7.20) the maximum permissible dose to which a pregnant woman shall be exposed is less than 1 rem (from the time she notifies her employer that she is pregnant). Under the Approved Code of Practice on Lead (see chapter 6) the limits of exposure to lead dust vapour and fumes is 40 micrograms per 100 millilitres of blood, and women of child-bearing age must have their blood tested on a regular basis. There are a number of statutory provisions which specify the maximum weights which women can lift (see para 7.39).

7.2 As a general rule, it can be stated that the critical exposure of a pregnant woman to deleterious substances is not so much in terms of the quantity or amount, but in the time of exposure. The first six weeks of pregnancy, when the foetus is being formed (and when possibly the woman might not even know she is pregnant) might well be the most critical time of all, together with the last three months of pregnancy, when the brain of the foetus is being formed. An American organisation – the National Institute of Occupational Safety and Health (NIOSH) – has issued guidelines which may usefully be followed when a woman is exposed to potentially harmful substances. These over (a) when a pregnant woman may continue to work, (b) when she may continue to work but with environmental modifications to accommodate her working, and (c) when she should not work at all. Similar guidance may shortly be available from the Society for Occupational Medicine.

7.3 There is a clear conflict between two schools of thought. On the one hand, there is the right of a woman to apply for employment and not to be discriminated against in the terms and conditions on which employment may be offered (Sex Discrimination Act, section 6), her right not to be dismissed because of pregnancy (Employment Protection (Consolidation) Act, section 60), her right to return to work after pregnancy (EPCA, section 45), and a further right in the child who may be subsequently born not to be exposed to any substance which may cause a congenital defect (see para 7.10). One the other hand, there is the duty of the employer to ensure the health and safety at work of all his employees, and as the duty is a personal one (see chapter 9) it must be fulfilled with respect to each woman who is of child-bearing age, not to expose her to the dangers (particularly from embryotoxic substances) which are risks to her health.

7.4 So far as discrimination is concerned, a recent decision of the EAT has thrown considerable light on the law. In *Page v Freight Hire (Tank Haulage) Ltd* the applicant drove a lorry carrying dimenthylformamide (DMF) from a chemical works to a storage plant. The employers were then informed by the makers (ICI Ltd) that there was an embryotoxic danger to women of child-bearing age if they came into contact with this chemical. Consequently, she was told that she could no longer be employed driving a vehicle containing this substance, and, as this was a casual job, she was deprived of work. She claimed that she had been discriminated against on grounds of her sex, but the argument was rejected. Section 51(1) of the Sex Discrimination Act states that the Act does not render unlawful any act done by a person in order to comply with the requirement of an Act of Parliament passed before the Sex Discrimination Act. The EAT was thus able to refer to HSWA (passed in 1974), section 2(2)(b) of which states the particular matters to which an employer should pay attention, i e 'arrangements for ensuring, so far as is reasonably practicable, safety and the absence of risks to health in connection with the use, handling, storage and transport of articles and substances'. Thus the employer's action was necessary to ensure compliance with the 1974 Act, section 51(1) of the Sex Discrimination Act applied, and there was no unlawful discrimination. It is important to note that health and safety considerations should not be used as a deliberate means of discriminating against women (or men, see *Peake v Automotive Products*), and the employer must be able to satisfy the tribunal that there was no other reasonable method of ensuring the health and safety of the person concerned.

7.5 Section 60 of the Employment Protection (Consolidation) Act 1978 provides that it will be unfair to dismiss a woman because she is pregnant (or for any other reason connected with her pregnancy) unless because of her pregnancy

(a) she is incapable of adequately doing the work she was employed to do, or
(b) she cannot do that work without contravention (either by her or by her employer) of a duty or restriction imposed by or under any enactment.

7.6 But even if an employer could justify the dismissal under one of these two headings, it will still be unfair to dismiss her unless he offers her another job where there is a suitable available vacancy.

7.7 Thus the fact that a woman is pregnant is by itself no grounds for dismissal. She must be incapable of doing the work, or there must be a legal restriction on her working, and in either case, the employer must search for some suitable alternative employment and offer it to her. For example, in *Brear v Wright Hudson*, an assistant who worked in a chemist's shop was dismissed when she became pregnant. The manager, who was a chemist with some medical knowledge, formed the view that she could no longer be employed because of her pregnancy, as the work involved lifting and carrying heavy boxes. Since, in a small shop, there was no other suitable alternative employment to offer, it was held that her dismissal was fair. However, in *Gamble v Tungston Batteries*, a different conclusion was reached. The applicant, who was pregnant, was unable to continue working on the manufacture of batteries because of the lead content. As there was no other suitable vacancy, she was dismissed. A fortnight later, a clerical worker left the firm, and the question arose as to whether or not the employers had acted unfairly in not offering this job to the applicant. It was held that as it was not 'available' at the time she was dismissed, a failure to offer it to her did not constitute unfairness on the part of the employers. However, the industrial tribunal thought that the employers had not made sufficient enquiries among the various departments to see whether or not there was an impending vacancy, and in failing to do this, they had acted unfairly in dismissing the applicant. Thus, if a pregnant woman was exposed to a suspected teratogen, it could be argued that the employers would be in breach of their duty under section 2 of HSWA in continuing to employ her, but before dismissing, suitable alternative employment should be considered.

7.8 To bring a claim under this heading, a woman must be employed for the requisite period of continuous employment of one year (*Singer v Millward Ladsky*) or two years in respect of employers who employ less than twenty employees (Employment Act 1980, section 8).

7.9 There should be few problems concerning the right to return to work after pregnancy, at least, so far as health and safety considerations are concerned, although it should be noted that it is unlawful for women to return to work in factories within four weeks of the childbirth (Public Health Act 1936, section 205; Factories Act 1961, Schedule 5).

7.10 It is generally accepted that English law would not provide a remedy for any injury suffered to a foetus while *en ventre sa mère*, there being no duty owed in tort, and no statutory duty applicable. The thalidomide tragedy highlighted this problem, and the Congenital Disabilities (Civil Liability) Act 1976 was passed. It was intended to be a temporary measure, pending legislation consequent on the report of the Pearson Commission, but as the recommendations of the latter are unlikely to be the subject of Parliamentary action in the foreseeable future, the Act remains. It provides that a child who is born disabled as a result of any breach of duty (whether through negligence or breach of statutory duty) to either parent, will be able to bring a civil action against the person responsible. In practice, this will mean that if something happens to a woman while she is pregnant which results in her child being born disabled, the child will have a separate right of action. Since the Limitation Act 1980 does not apply to a minor (i e one under the age of eighteen) he may be able to bring a claim any time within three years from attaining his majority, i e potentially twenty-one years after the event which caused the disability! There must be a breach of duty to the child's parent, but it is not necessary that the parent be injured. The disability may be something which happened to either parent which prevents the mother from having a normal child, or an injury to the mother while she is pregnant, or an injury to the foetus during pregnancy.

7.11 If the person responsible would have been liable to the parent, he is liable to the child. The child must be born alive, for there is no duty to the foetus as such. It is not relevant that the person responsible knew that the woman was pregnant, as long as a duty was owed to her. Thus if a woman is working with a chemical which causes her to have a disabled child as a result of ingesting minute particles, an action will lie against the employer by the child.

7.12 This can cause some concern among those employers who regularly use chemicals, for there are a number of substances which are suspect so far as pregnant women are concerned, e g vinyl chloride, benzene, mercury, anathetic fumes, lead, etc. Molecules of almost any substance can pass through the placenta, and therefore any woman may, by ingesting, inhaling, or through skin absorption, pass a deleterious ingredient into her body.

7.13 The Act poses problems for employers who have to balance delicately between the law against sex discrimination on the one hand, and liability to unknown potential plaintiffs on the other. The decision in *Page v Freight Hire (Tank Haulage)* (above) may be of some assistance, provided the employer acts with the fullest information and with a genuine concern for health and safety.

Mental illness

7.14 This term encompasses a wide range of medical and psychological problems, but from the point of view of health and safety, an employer has to consider (a) whether to employ someone who has, or who has had, mental illness, (b) if he is employed, what steps should be taken to ensure that he is not a health or safety hazard to himself or to other, and (c) whether or not a past or present illness is sufficient ground for dismissal.

7.15 There are no relevant statutory provisions to give guidance, and it is not unlawful to refuse to employ someone who is suffering from, or who has suffered from, mental illness. The problem is one of medical evidence, and recruitment should be done in consultation with such specialists as are available (e g company doctor, nurse, etc), so that a full and satisfactory assessment can be made of the illness and the risks involved in employing the applicant. Generally, mental illnesses are of two types, namely psychoneurosis and psychosis. The former is a species of 'bad nerves', the latter is more serious, and involves a lack of contact with reality. So far as psychoneurosis is concerned, recovery rate is good, and a person who has had a previous history of this sort of mental illness, but who is now fit and well, is no more likely to have a recurrence of the illness than a person who has no previous history is likely to have a mental breakdown. The prognosis for psychosis, however, is mixed, depending on the severity of the illness and the medical evidence which can be adduced.

7.16 Many companies, while they may be unwilling to take on persons with a history of mental illness, adopt a policy of social responsibility to existing employees, and seek to provide suitable employment, rehabilitation units, and encourage the employee to obtain medical treatment. Satisfying work, of course, is a great therapy. The problem is one which requires sympathetic handling, involving the co-operation of all those who can contribute some expertise, as well as employees and management generally. Progress should be monitored, risks assessed, work planned and adequately supervised, and so on. Once the nature of the illness can be established, it may be possible to match it with employment where safety can be observed.

7.17 The fact that an employee 'tells a lie' on a job application form and denies that he has ever had any history of mental illness is not, *per se* grounds for dismissal should the previous history come to light (*Johnson v Tesco Stores*), but if it can be shown that there is a good sound, functional reason why the person cannot be employed, it may be fair to dismiss him. In *O'Brien v Prudential Assurance Co*, when the applicant was interviewed for a job, he was given a medical examination, and asked specific questions about his mental health. He made no mention of the fact that he had had a long history of mental illness, including psychosis, which necessitated the taking of drugs and a period of hospitalisation. When the truth can to light, the company consulted various medical experts, and a decision was taken to dismiss him. He brought a claim for unfair dismissal, but it was held that the dismissal was fair for some other substantial reason. It was the company's policy not to appoint as district agents persons who had long histories of mental illnesses, as the job involved going into peoples' homes. Clearly, there was a risk that an unpleasant incident could occur which would seriously tarnish the company's image, and thus the policy was not an unreasonable one. Had the company been aware of his long history of mental illness he would never have been appointed to the job.

7.18 The greater the risk of a serious incident occurring, the more management should consider dismissing the employee. In *Singh-Deu v Chloride Metals* the applicant was sent home from work after complaining that he felt unwell. His doctor diagnosed paranoid schizophrenia, and when the company could not obtain a satisfactory assurance from a specialist that there would not be a recurrence, he was dismissed. This was held to be fair. If he had an attack of this illness during working hours, a catastrophic accident could have happened. There was no other suitable alternative employment for him, and the company could not be expected to wait until an accident occurred before taking some action (see *Spalding v Port of London Authority*).

Radiation

7.19 The more dangerous type of ionising radiation comes from X-rays and other radio-active material, and non-ionising radiation can stem from laser beams, arc-welders, infra-red and ultra violet sources, microwave ovens, etc. The effect of radiation exposure will vary according to the dose and exposure time, but it is generally accepted that serious illnesses and even death can result, as well as damage being caused to a person's genetic structure, causing still-births and malformation in newly born children. The exposure limits at present adopted in this country are those recommended by the International Commission on Radiological Protection, and these limits vary with the area of the body exposed. However, there appears to be no valid proof that the thresholds recommended are in fact safe, and any radiation exposure must therefore be regarded as being potentially damaging. Proper shielding is therefore essential (principally by the use of lead material), the equipment must be properly maintained so as to prevent leakage, radiation film badges should be worn at all times when there is a likelihood of exposure in order to measure the amount (or a dosemeter used), and whole body counts should be made when there is a danger that an affected person may have absorbed or ingested radiation. The National Radiological Protection Board has issued a guidance booklet 'Protection against Ultra Violet Radiation in the Workplace' which should be consulted, and a number of other advisory publications can be obtained from HSE.

7.20 The Ionising Radiations (Sealed Sources) Regulations 1969 and the Ionising Radiation (Unsealed Radioactive Substances) Regulations 1968 lay down the maximum permissible doses to which a person employed in a factory may be exposed in each calendar year. These are as follows: 75 rems in the hands, forearms, feet and ankles, of which not more than 40 rems shall be received in any calender quarter; 15 rems to the lens of the eye, of which not more than 8 rems shall be received in any calender quarter; 30 rems to any other part of the body, of which not more than 15 rems shall be received in any calender quarter. There are even more stringent limits relating to an exposure to X-rays, gamma rays and neutrons.

7.21 If an inspector has reasonable cause to believe that a person employed in a factory has received in any calendar year or calendar quarter a radiation dose greater than 3/10th of the permitted maximum dose, he may serve a written notice on the occupier requiring him to make arrangements for the protection of the person employed, including the provision and wearing of film badges or dosemeters, periodic medical examination, suspension from work involving exposure to ionising radiations, the monitoring of the factory, and for keeping and preserving records of measurements obtained by such monitoring.

7.22 The Regulations also make provision for the marking off of radiation area boundaries, the provision of film badges or dosemeters, constant medical supervision, adequate training and instruction concerning the hazards involved and on the precautions to be observed, and the handling of sealed and unsealed sources and substances. If a factory owner knows, or has reason to believe, that a female employee is pregnant, she must not be employed by him on any work where she may be exposed to X-rays, gamma rays, or neutrons, in excess of 1 rem.

7.23 The Regulations also contain detailed rules to be observed by employees who are at risk from exposure. They must report any suspected excessive dose, submit to medical examinations (including blood sampling), wear the personal protective equipment provided, remove the equipment on leaving active areas, depositing it in a proper place, wash hands and test for contamination, remove any contamination and report residual excessive contamination, refrain from drinking, eating, or smoking in active areas, go for first aid promptly if skin is cut or broken, report any accidental spill or escape, refrain from entering contaminated areas without written permission, and refrain from wearing personal clothing which is contaminated to an excessive degree.

7.24 Radiation which results from nuclear energy sources is controlled by the Nuclear Installations (Dangerous Occurrences)

Regulations 1965, which are enforced by the Nuclear Inspectorate of HSE. By the Radioactive Substances Act 1960, no person may keep or use any radioactive materials on premises used for the purpose of an undertaking carried on by him unless he is registered with the Department of the Environment or Scottish Office or otherwise exempted.

7.25 The European Commission has made proposals for the revision of the Euratom Directive of 1976 on protection against radiological hazards, and new Regulations giving effect to these are expected shortly (see chapter 12).

Threshold Limit Values (TLVs)

7.26 This term refers to the allowable airborne concentrations of harmful substances which may enter the human body through inhalation, ingestion or swallowing, or absorption through the skin or linings of the mouth or nose. The airborne contaminants are measured as follows: in the case of fumes, vapours or gases, the levels are determined by the parts found per million of air (ppm) or alternatively, milligrammes per cubit metre of air (mg/M^3). In the case of dust, it will be measured in mg/M^3, or alternatively millions of particles per cubic foot of air (mppcf).

7.27 There is no hard and fast dividing line between safe and harmful contaminants, as a number of factors must be taken into account, including working conditions, individual susceptibility, length of exposure, the presence of other substances which may increase or decrease the toxicity, and so on. Further, while some substances may be harmless, they may have a nuisance value, affect concentration or cause stress, as well as militating against pleasant working conditions.

7.28 TLV's can be measured in three ways.

7.29 (a) *Time Weighted Average (TWA)* This is the concentration to be found based on a normal working day of eight hours, or a normal working week of forty hours. Thus exposure above the appropriate limits may be compensated for by a decrease in exposure for some of the time.

7.30 (b) *Short Term Exposure Limits (STEL)* This is the maximum limit to which a person should be exposed for a period of fifteen minutes. No more than four such exposures are permitted in any day, and there must be at least sixty minutes between the periods of exposure.

7.31 (c) *Ceiling (C)* This is the limit which must not be exceeded at any moment in time, and is usually applicable to substances which are fast acting in their adverse biological effect.

7.32 Although there is no specific legislation covering LTVs, *per se*, there are statutory provisions relating to fumes (e g FA, section 63) and precautions which must be taken when certain chemicals are present by requiring the provision of masks, etc (Agricultural (Poisonous Substances) Regulations), or to prevent ingestion by other means (Control of Lead at Work Regulations). Thus once the problem is known to exist, appropriate measures should be taken, safe TLVs established, regular monitoring procedures adopted, and protective equipment provided.

Heavy loads

7.33 It is estimated that about 20,000,000 working days are lost each year as a result of workpeople suffering from back pain, which causes some 80,000 employees to be absent from work each day. Back pain is likely to be more common in those industries which require the lifting of heavy loads, but this need not be exclusively so, and anyone who has to lift a parcel or package of moderate weight or who is constantly bending and straightening his trunk is at risk. Poor posture while sitting and certain medical disorders can also bring about a fair number of complaints.

7.34 The main protective measures which can be taken include:

(a) training employees in proper lifting and handling techniques. A number of organisations run courses on this topic which should bring about a dramatic reduction in the accident rate;
(b) the introduction of mechanical lifting and handling machinery;
(c) screening examinations, in order to discover those employees who are prone to this kind of accident. However, this will only result in the elimination of a few people who are at risk, for the majority of persons who suffer back pain are likely to be 'first timers';
(d) redesign of job operations and seating accommodation so as to improve posture or eliminate risk.

7.35 Advice generally on this problem can be obtained from the Back Pain Association, Grundy House, Teddington, Middlesex.

7.36 The statutory provisions which are relevant are somewhat general. The Factories Act, section 72 states that a person shall not be employed to lift, carry or move any load so heavy as to be likely to cause injury (see chapter 4) and similar provisions can be found in OSRPA, section 23 (see chapter 5), and the Agriculture (Safety, Health and Welfare Provisions) Act 1956, section 2 (see chapter 8). The problem is complicated by the fact that various individual have different weight-lifting capacities which will depend on their strength, training, age, experience and susceptibilities to back strain, as well as the nature of the load to be carried, which may be rigid or compact. The result is that the legislation (apart from certain Regulations, see below) contains only statements of general principles, and each case must be regarded on its merits.

7.37 Whether an employee is employed to move a heavy weight may depend on whether or not there was no other way in which he could do the work. In *Brown v Allied Ironfounders Ltd* the plaintiff was informed that if she needed help to move a stilage, she should ask for assistance. This was normal practice, but it was also known to the foreman that some women moved the stilages by themselves, and no express prohibition was issued. The plaintiff was injured when she tried to move a heavy stilage, and it was held that the employers were liable for the injury. To leave it to her to decide whether or not to ask for assistance meant that the moving of the load was part of her job, which she was employed to do. A somewhat different conclusion was reached in *Black v Carricks (Caterers) Ltd* (see chapter 5) and the lines between these and other cases must be regarded as extremely fine.

7.38 Whether a load is likely to cause injury is also a factual issue. In *Kinsella v Harris Lebus Ltd* it was held that a load of 145 lbs was not so heavy as to be likely to cause injury to a man of experience, and in *Peat v Muschamp & Co Ltd* a weight of 65 lbs was not too heavy for a one-legged man to lift, as he regularly lifted weights of that amount, and was told that he could ask for assistance if he needed it. It is a fallacy to assume that the sharing of a load will result in the equalisation of the weight to be carried. In *Fricker v Benjamin Perry & Sons Ltd* four dockers were carrying a crate of tobacco which weighed 560 lbs. It may be assumed that the equal weight between them would mean that each carried 140 lbs, but when one of the dockers was injured as a result of this operation, the employers were held liable.

7.39 The following legislative provisions specify the maximum weights which may be lifted:

Jute (Safety, Health and Welfare) Regulations 1948
7.40

	MAXIMUM WEIGHT, COMPACT OR RIGID BODY	MAXIMUM WEIGHT, NOT COMPACT OR RIGID
Women aged 18 and over	65 lbs	50 lbs
Young males aged between 16–18	65	50
Females aged between 16–18	50	40
Young person under 16	40	35

Pottery (Health and Welfare) Special Regulations 1950
7.41 Women must not be allowed to lift saggers if the weight (plus its contents) exceed 30 lbs, or to lift in conjunction with another person if the sagger exceeds 50 lbs. However, if the move is no more than 6 feet, these weights can be increased to 50 and 80 lbs respectively.

7.42 Young persons must not life weights exceeding 20 lbs unless an Employment Medical Adviser or appointed doctor gives a certificate stating a higher maximum weight may be carried.

Woollen and Worsted-Textiles (Lifting of Heavy Weights) Regulations 1926
7.43

	MAXIMUM WEIGHT, RIGID OR COMPACT	MAXIMUM WEIGHT, NOT RIGID OR COMPACT
Adult male	150	120
Woman aged 18 and over	65	50
Male young person (aged 16–18)	65	50
Female young person (aged under 18)	50	40
Male under 16	50	40

Agriculture (Lifting of Heavy Weights) Regulations 1959
7.44 The maximum weight of any sack or bag which may be lifted unaided by a worker employed in agriculture is 180 lbs.

Occupational deafness

7.45 It appears to be generally accepted that excessive exposure to noise levels in excess of 90dB(A) will cause permanent damage to a person's hearing, as will exposure to louder impact noises (explosions, hammer guns, etc). This does not mean that every deaf person has been injured because of the hazards of his occupation, for there are several other causes of deafness, including presbyacusis, which is brought on by simple old age. Since noise is a common feature of everyday living, it cannot be assumed that occupational noise is the sole cause of deafness; one wonders, for example, about the effect on the hearing of young people who regularly attend modern discos. However, the hazards to the hearing of workers in certain industries have been well known and documented for many years (boilermakers etc), and if a noise level reaches 90dB(A) exposure time should be reduced, and protective measures taken.

7.46 There is no general legislative provision relating specifically to noise. Section 2(2)(e) of HSWA requires an employer to provide and maintain a working environment that is, so far as is reasonably practicable, safe and without risks to health, and it is submitted that this provision is wide enough to encompass excessive noise in an occupation. There appears to be no reason why an inspector could not issue an Improvement Notice requiring an employer to introduce sound reduction techniques if the circumstances were appropriate. The Woodworking Machines Regulations 1974 provide that in every factory mainly used for work carried on by woodworking machines, if any person is likely to be exposed continuously to a sound level of 90dB(A), reasonably practicable measures shall be taken to reduce noise, and suitable ear protectors shall be provided and made readily available, and these shall be used by persons who are exposed to the noise. The Agriculture (Tractor Cabs) Regulations 1974 provide that safety cabs must be constructed so as to afford protection to the driver, including protection from noise, for which it was designed. Before a certificate of approval will be issued, it must be shown that the noise levels do not exceed 90dB(A). The Offshore Installations (Construction and Survey) Regulations 1974 and the Offshore Installations (Operational Safety, Health and Welfare) Regulations 1976 also deal with certain noise prevention requirements.

7.47 The Department of Employment has issued a Code of

Practice on 'Reducing the Exposure of Employed Persons to Noise'. This deals with the environmental control of noise, the design and maintenance of machinery, ear protectors, audiometric testing, persuasion techniques, etc. HSE has issued a booklet 'Noise and the Worker' which may also be consulted.

7.48 The problem of noise at work can be tackled in four ways. First, by reducing the noise at its source. This involves new design techniques, new technology, new materials, etc. Consultations can take place with manufacturers (reminding them of their obligations under section 6 of HSWA). The level of 90dB(A) is a somewhat arbitrary level which it was thought could be afforded by industry, bearing in mind the cost of replacing existing machinery, but it is not a safe standard in absolute terms, and ideally, a level not exceeding 85dB(A) should be aimed at.

7.49 Second, noise reduction techniques can be implemented. This may involve isolating the source of the noise, so that work-people are unaffected by it, or blanketing walls and ceilings with noise absorbing materials, which will have the effect of reducing noise levels, and so on.

7.50 Third, ear protection should be provided, appropriate to the dangers present. This can be achieved by ear plugs, ear muffs, helmets, etc, but it is important that the correct protection is provided in each case, depending on the level of noise present. The object must be to reduce harmful sound, not sound itself, for it is essential to retain the latter (e g as a warning of danger, or to receive instructions, etc). Ear plugs should be fitted by a qualified person (occupational nurse, etc) otherwise the selection of a wrong size may render the protection ineffective. Ear muffs must be comfortable to wear, aesthetically acceptable to the wearer, and must not interfere with the work or other protective equipment.

7.51 Fourth, exposure time to excessive noise can be reduced by allowing rest periods in rest rooms away from the noise, or operating a rota system.

7.52 The main legal remedy in respect of occupational deafness is a claim for compensation based on common law negligence (see *Berry v Stone Manganese*). The existence of a duty of care is now well recognised, and while a failure to provide the necessary precautions will clearly amount to a breach of that duty, there is some room for argument as to the duty to compel, exhort or pro-pagandise in order to ensure that they are used. An employee may not readily realise the insidious nature of the danger of working without the precautions or the fact that he may suffer permanent damage. Something more positive is required from the employer than the passive duty to provide ear muffs, plugs, etc. It also follows that if an employee fails to use the precautions after being instructed in their proper use, any compensation awarded can be reduced in respect of his contributory conduct. Compensation awards have been awarded in the region of £4,000–8,000, depending on the extent of the deafness (see *Smith v British Rail Engineering Ltd*). Any claim must be brought within three years from the damage (Limitations Act 1980), but since deafness is a process which takes place over the years, an affected person may not know in the early stages of the nature or extent of the damage caused.

7.53 The Social Security (Industrial Injuries) (Prescribed Diseases) Regulations 1975 (as amended) enable a worker to obtain industrial injuries disablement benefit (see chapter 9) if he suffers from occupational deafness. The worker must have been employed in the prescribed occupation (see Appendix) for more than twenty years (and in at least one year before the claim), and he must suffer from an average hearing loss of at least 50 decibels in both ears. In at least one ear, the loss must be due to noise at work. HSE has recently issued a Consultative Document entitled 'Protection of Hearing at Work' which outlines proposals for new Regulations, an Approved Code of Practice and Guidance Notes, and these are currently under discussion (see chapter 12).

Visual display units

7.54 It seems to be accepted that there is no likelihood of there being any serious occupational hazard from the use of visual display units, although good management can reduce to a minimum certain discomforts which may produce adverse results from the point of view of efficiency. The main problem is likely to be eye-strain, which is a fatigue in the eye muscles. Preventative measures here might include regular eye tests, the provision of suitable spectacles (not bi-focals) and keeping the equipment in good order. Attention should also be paid to the seating and lighting arrangements, and proper rest periods should be allowed for. It has been suggested that employees who suffer from epilepsy, migraine or some nervous complaints should not be required to work with VDU's, as there appears to be a possibility that an attack may be brought on.

Working alone

7.55 There are some situations where there is a serious risk to a person if he is injured while working alone, because he is unable to summon help. Three specific Regulations therefore make working alone an offence, namely Chemical Works Regulations 1922, Electricity (Factories Act) (Special Regulations) 1944, and Work in Compressed Air (Special Regulations) 1958. The problem here is how to avoid overmanning, yet provide adequate back-up staff in case of accidents. The CBI has published a guidance document which should be looked at in appropriate cases.

Immigrant workers

7.56 There is no evidence to suggest that immigrant workers are more prone to accidents than any other group of employees, although since immigrants tend to be employed in those industries where there are greater dangers, obviously they may have a higher than average accident rate. The problem must be tackled by means of adequate training, and language and cultural factors must be taken into account. Safety instructions may have to be prepared in the language of the worker concerned, and signs and posters which need little translation should be used to indicate the hazards and/or the precautions to be taken. In *James v Hepworth and Grandage Ltd* the employers put up a notice which stated that spats should be worn. Unknown to them, the plaintiff could not read English, and he was injured through not wearing spats. His claim for compensation failed. He had observed other workmen wearing spats, and his failure to make any enquiries about them led to the conclusion that he would not have worn them anyway. However, this case cannot be regarded as authority for the proposition that an employer fulfils his legal duty by merely drawing attention to a precaution (see chapter 9) and if there is a risk of a more serious injury, greater steps must be taken to ensure the use of protective clothing etc. Further, in *Hawkins v Ian Ross (Castings) Ltd* an employee was injured partly because he was working alongside an immigrant who had a limited command of English, and who misunderstood a warning shout. The court held that in such circumstances, a higher standard of care is required on the employers when considering the layout of the work and the steps to be taken to avoid accidents.

7.57 Language barriers may not be used as a device to exclude immigrants from employment, and attempts to impose language tests have resulted in allegations of racial discrimination. The problem will no doubt be with us for many years, for although we tend to equate immigration with the influx from Asian or

African countries, under article 48 of the Treaty of Rome, the free movement of workers throughout the European Communities is guaranteed, and an influx of non-English speaking workers from Europe cannot be excluded. Training for safety officers may well include some form of instruction in foreign languages in the not-too-distant future!

Epileptics

7.58 Reconciling safety with social responsibility is one of the more difficult tasks of recruitment officers, and it is essential that judgments be made on the basis of informed opinion. This is particularly true of epilepsy, about which there is a great deal of ignorance. There is no reason in principle why an epileptic cannot be employed in most types of work, provided a full appraisal is made of the situation. Epilepsy does not conform to a single pattern; it varies in type, frequency, severity and timing. A candidate for a job who has all the necessary qualifications and experience should not be excluded from consideration on the ground of epilepsy without a full investigation, in which, of course, medical assessment would be invaluable.

7.59 A person who has a history of epilepsy cannot be permitted to drive a heavy goods vehicle or passenger vehicle, or work as a specialist teacher of physical education. Other than these restrictions, objections to appointing an epileptic should be based on sound functional reasons. Once employed, safety considerations obviously become important, and if there is any risk to the employee or to his fellow employees, attempts must be made to find suitable employment where the risks are minimised. In the last resort, however, firmer action must be taken. For example, in *Harper v National Coal Board* the applicant had three epileptic fits in a period of two years. During these fits, quite unknowingly, he was violent to other workers. He was then medically examined, and it was recommended that he be retired on medical grounds, under a miner-workers pension scheme. He refused to accept retirement, and was dismissed. This was held to be fair. The employers owed a duty to their other employees to ensure their safety as well. The applicant was working alongside other disabled employees, and they were clearly at risk.

7.60 Advice on the employment of epileptics can be obtained from the British Epileptic Association, and the Employment Medical Advisory Service may also be contacted when necessary.

Disabled employees

7.61 Under the Disabled Persons (Employment) Act 1958 every employer must employ at least 3 percent of his workforce who are registered disabled persons. Certain occupations, such as car park and lift attendants are designated as being particularly suitable for disabled persons, and a person who does not hold a 'green card' may not be employed without giving first consideration to a disabled person. Unfortunately, many of the jobs which used to be suitable for disabled employees are beginning to disappear, and it now appears that more than 60 percent of all employers have failed to reach their quota. The problem arises partly because there are not enough registered disabled applicants who have the right qualifications for the vacant jobs, and partly because many employers are reluctant to take on disabled applicants due to possible safety hazards. Although it is an offence not to employ the quota of disabled persons, a prosecution may only be brought with the consent of a special advisory committee, for the problem is one which must be solved by persuasion rather than by compulsion.

7.62 Under the Companies (Directors Reports) (Employment of Disabled Persons) Regulations 1980, every director's report made under the provisions of the Companies Act relating to a financial year shall contain a statement describing the policy of the company throughout that year for

(a) giving full and fair consideration to applications for employment made by disabled persons, having regard to their particular aptitude and abilities;
(b) continuing the employment of, and the arranging of appropriate training for, employees of the company who have become disabled persons during the period when they were employed by the company; and
(c) the training, career development and promotion of disabled persons employed by the company.

7.63 The Regulations apply to every company which employs on average more than 250 employees throughout the year in question. It is hoped that the new legal requirements, by throwing a spotlight of publicity on the problem, will encourage more companies to provide job opportunities for disabled persons.

7.64 Once a disabled person has been taken into employment, he can, of course, be dismissed on health or safety grounds, but there is a high burden on the employer to show that he made a full appraisal of all the facts, and in particular that he searched around for suitable alternative employment before contemplating dismissal. The employer must familiarise himself with the job, the disability and the hazard (*Littlewood v AEI Cables*), and extra efforts must be made to accommodate a disabled person. In *Cannon v Scandecor*, when the applicant, who was registered disabled, was employed, it was recognised that her output would not be too high. She then had a series of illnesses and a car accident, and when she returned to work, her daily output was that which she had previously achieved in an hour. She was dismissed, but this was held to be unfair. Much too short a period had elapsed between her return to work and her dismissal, and sufficient time had not been allowed in order to enable her to make a full recovery. She should have been given a reasonable trial period, and then, if her output had not reached her previous performance, her dismissal would have been fair.

7.65 It will thus be evident that employers who take on a disabled person have a special responsibility to do more than they would normally do to give him a period of time in which to reach acceptable standards. Some companies meet the problem by excluding disabled employees from departmental budgets, so as to ensure that they are not regarded as a drain or millstone by management seeking to reach targets. Certainly, a great deal can be done, and although safety hazards must not be ignored, these are not always insuperable obstacles. There is no substitute for a careful asssessment of the nature of the work, the risks associated with it, and a measured judgment based on an investigation into the nature of the disability matched against the risks involved.

7.66 Even so, there must come a point in time when safety factors will override all other considerations, although the duty of the employer is to act reasonably when handling employees who have any type of disability. Usually this will mean taking medical advice, and looking round for suitable employment. In *McCall v Post Office*, the applicant, unknown to his employers, suffered from epilepsy. He was employed as a cleaner, but had several fits which were of short duration. The personnel department took medical advice and were told that he should be employed in a 'no risk' area. Unfortunately, there was no suitable vacancy, and he was dismissed. This was held to be fair. On medical advice, the employee had to be moved to a safe area, and as there was none available, it was in his interests, and in the interests of other employees, that he be dismissed.

8 Agriculture and construction

A Agriculture

8.1 Nearly 100 people employed in agriculture are likely to be killed each year, a quarter of whom may be under the age of 16, and there are over 4,000 non-fatal accidents. Agricultural workers frequently work alone in remote places where immediate attention is not readily available. Supervision is minimal. The industry is highly mechanised, yet the general working environment is far from ideal, and many dangerous substances are used. These facts illustrate the need for strong protective legislative measures and a greater need for educating workers in safe practices and procedures. The annual report from HSE's Agricultural Inspectorate discusses many of the hazards which arise whilst working on farms, and considers some of the special problems. The Agricultural Industry Advisory Committee is currently engaged in a wide programme of work in looking at improvements which can be made to increase safety in the industry generally. A new permanent office and Health and Safety Information Centre has been established at Stoneleigh in Warwickshire, which aims to provide up-to-the minute advice and information on a wide range of health and safety problems.

8.2 One of the recommendations of the Gowers Committee which reported in 1949 was that there should be legislation to cover safety hazards which arise out of employment in agriculture. In this chapter we will be concerned with those legislative provisions which are still in force, and which are therefore in addition to the general law.

Agriculture (Safety, Health and Welfare Provisions) Act 1956

8.3 This Act was designed to provide for the health, safety and welfare of persons employed in agriculture, and the avoidance of accidents to children arising out of the use of vehicles, machinery or implements used in connection with agriculture.

Lifting heavy weights (section 2)

8.4 A young person (i e a person who is over compulsory school-leaving age, but who has not yet reached the age of eighteen) shall not be employed as a worker in agriculture to lift, carry or move a load so heavy as to be likely to cause injury to him. By the provisions of the Agriculture (Lifting of Heavy Weights) Regulations 1959, the maximum weight which may be carried unaided by any worker employed in agriculture in a load which consists of a sack or bag shall not exceed 180 lbs.

Sanitary conveniences and washing facilities (section 3)

8.5 If HSE considers that any agricultural unit is without suitable and sufficient sanitary conveniences, or suitable and sufficient washing facilities, the inspector may serve a Notice on the appropriate person (which may be the landlord or occupier, as the circumstances dictate) requiring him within the specified time to execute such works or take other such steps as may be specified for the purpose of providing suitable and sufficient conveniences or facilities. In making this judgment, the inspector shall have regard to the number and sex of the persons employed, the location and duration of the work, and all relevant circumstances. In addition to a Notice under this section, there is also the power of the inspector to issue an Improvement Notice under HSWA. One distinction between these two procedures is that under HSWA, appeals against an Improvement Notice must go to the industrial tribunals, whereas under the Agriculture (Safety etc Provisions) Act an appeal will lie to the magistrates' court within twenty-eight days, with a further appeal to the Crown Court.

8.6 Facilities cannot be regarded as being suitable if they are some distance away. For example, in *Fairhurst v Luke*, an Improvement Notice was served requiring a farmer to provide accessible toilet facilities. To reach the nearest lavatories, it would have taken workmen four-and-a-half minutes by tractor, eight minutes by bicycle, or ninety seconds in a car. It was held by the industrial tribunal that in view of the isolated nature of the farm, the Improvement Notice would be upheld, and the farmer was ordered to provide additional facilities. Otherwise, it was quite clear that rather than make such a long journey, farm workers would tend to relieve themselves on the spot, and this would create a health hazard (one may query whether this would create any greater risk than that which stemmed from animals, but the provision of additional facilities would certainly make body functions more comfortable).

Cleanliness of sanitary conveniences (section 5)

8.7 If an HSE inspector considers that sanitary conveniences are not being properly maintained or kept clean he shall serve a Notice requiring the occupier of the unit to take such steps as may be specified to secure the proper maintenance or to cleanse it.

First aid

8.8 The provisions of the Act relating to first aid have been repealed and replaced by the Health and Safety (First-Aid) Regulations 1981 (see chapter 6).

Avoidance of accidents to children (section 7)

8.9 It is an offence to cause or permit a child to ride on or to drive a vehicle or machine or agricultural implement in contravention of the Agriculture (Avoidance of Accidents to Children) Regulations 1958. These provide that no child (i e a person below the age of thirteen) shall ride on certain agricultural vehicles while they are being used in the course of agricultural operations. The vehicles in question are tractors, self-propelled agricultural machines, trailers, trailers into which a conveyor mechanism is built, machines mounted on or towed by tractors, binders or mowers drawn by animals. A child may ride on the floor of a trailer, or on any load carried by a trailer, provided it has four sides each of which is higher than the load. A child shall not drive a tractor or self-propelled vehicle or machine while it is being used in the course of agricultural operations, or ride on agricultural implements.

Regulations relating to agriculture

8.10 There are a number of Regulations which relate to agriculture, and the following is a summary of the relevant provisions.

AGRICULTURE (LADDERS) REGULATIONS 1957

8.11 These require every employer to ensure that employees do not use ladders in the course of their employment unless the ladders are of good construction and sound material, and properly maintained. Neither the employer nor the employee shall use a ladder if it is not strong enough for the purpose and manner of use, or if any rung is missing. Every employee shall report any defects in ladders to his employer.

AGRICULTURAL (POWER TAKE-OFF) REGULATIONS 1957

8.12 These provide that an employer shall not cause or permit a worker to use any tractor with a power take-off while the engine is in motion, unless it is guarded by a shield to protect the worker (or his clothes) from coming into contact with the power take-off, or it is not in use and is completely enclosed by a cover. The entire length of the take-off shaft shall be wholly enclosed by a guard while in motion.

AGRICULTURE (CIRCULAR SAWS) REGULATIONS 1959

8.13 These Regulations lay down provisions in relation to agriculture which are similar to the Woodworking Machine Regulations 1974. Part I of Schedule 1 to the Regulations lays down obligations on employers to ensure that circular saws shall be substantially constructed and properly maintained. Adequate artificial or natural light shall be provided when circular saws are being used, and defective blades must be replaced. Top guards which are rigid shall be provided, and bottom guards installed below the bench table. There are special provisions for circular saws with swing tables.

8.14 Part II of Schedule 1 places obligations on workers. They must keep in position and make use of riving knives and guards, and report any defect to their employer. Both employer and worker must ensure that the floor around a circular saw is unobstructed, and push sticks to push blocks shall be used whenever a risk of injury would be reduced. No adjustment shall be made while the saw blade is in motion. Schedule 3 of the Regulations provides that a worker shall not operate a circular saw unless its workings have been demonstrated by an experienced person who is over the age of eighteen. A worker who is over the age of sixteen but below the age of eighteen shall not use a circular saw except under the supervision of a person who is over the age of eighteen and who has a knowledge of how it operates. A worker under the age of sixteen is prohibited from using a circular saw altogether.

AGRICULTURE (SAFEGUARDING OF WORKPLACES) REGULATIONS 1959

8.15 These provide that floors and stairways shall be as safe as is reasonably practicable. There are special requirements in relation to handrails for stairways and secure handholds for steep stairways. Apertures in floors and walls must be guarded by a cover or fence or guardrail, as must grain pits, stokeholds and furnace pits. Defects in steps, handrails, covers, fences and guardrails which are discovered by a worker must be reported to his employer.

AGRICULTURE (STATIONARY MACHINERY) REGULATIONS 1959

8.16 These lay down the precautions which must be taken in respect of stationary machines. Every component shall be so situated or guarded as to protect a worker from coming into contact with it, and similar requirements are laid down with respect to primary driving belts, feeding inlets and discharge outlets. Prime movers and stationary machinery must have a device by which they can be quickly stopped, belts must be properly maintained, and there must be adequate natural or artificial lighting provided. Workers must use safety appliances, and damaged guards must be reported to the employer.

AGRICULTURE (THRESHERS AND BALERS) REGULATIONS 1960

8.17 These state that the drum feeding mouth of a thresher shall be guarded to prevent a worker from coming into contact with the drum, or covered, as appropriate. The deck of every thresher shall be fitted with a guard rail, or rope, chain or fence if a worker is liable to fall more than 5 feet. Pointed hooks and spikes must not be used for the attachment of a sack or bag to the thresher. Balers and trussers shall also be guarded. Threshers and balers shall be of sound construction and properly maintained, components shall be adequately guarded, provision must be made for a device which will stop the machinery quickly, and there shall be adequate natural or artificial light provided when these machines are being used. Workers shall use all safety precautions, and damaged guards must be reported to the employer.

AGRICULTURE (FIELD MACHINERY) REGULATIONS 1962

8.18 These Regulations apply to those machines which are used for agriculture other than a machine which is stationary (excluding self-propelled vehicles which are primarily designed to carry persons or loads, and aircraft). All components driven by any ground wheel or which are driven by power must be guarded, or situated so as to prevent the worker from coming into contact with them. If the machine has rotating knives, tines, flails etc (excluding cylinder mowers, hedge cutters, etc) the guard must cover the operative parts as near as is practicable, and if the parts rotate, the guard must cover the top and as near to the ground as is possible on a vertical rotation, and extend at least 1 foot beyond the periphery where the rotation is horizontal. There are provisions which relate to the guarding of power driven potato spinners, chains saws, rotary hedge cutters, and pick-up balers. Stopping devices shall be fitted on prime movers, and differential locks shall indicate to the driver whether or not the gear is locked. Valves and cocks shall indicate the effect of movement, draw-bar jacks shall be fitted, standing platforms shall have toeboards and guardrails. If the field machine has seats, these must be of adequate strength, with a back-rest and foot-rest, and mounting devices shall be provided to enable a worker to mount or dismount safely. All field machines shall be maintained in a safe condition, and safety devices shall be of adequate strength. Workers shall use safety devices, and report any defect to their employer.

AGRICULTURE (TRACTOR CABS) REGULATIONS 1974

8.19 These enable HSE to approve a safety cab for use with a tractor by issuing a certificate of approval. HSE must be satisfied that the model conforms to BS no 4063 (1963), and that the noise levels inside the cab do not exceed 90dB(A) if tested in accordance with BS no 4063 (1973). It is a condition of approval that the manufacturer to whom the approval is issued shall ensure that the approval mark is on every cab sold or let on hire, and he must make available for inspection any safety cab manufactured, and submit to any tests. By the Agriculture (Tractor Cabs) (Amendment) Regulations 1980 tractor cabs of a type approved by the European Communities are acceptable as an alternative to British Standards, and an EC mark is deemed to satisfy the Regulations.

8.20 A person shall not sell a new tractor or let one on hire, unless it is properly fitted with an approved safety cab, and is marked with the appropriate approval mark. Every employer shall ensure that every tractor driven by a worker in the course of his employment is properly fitted with a safety cab marked with the appropriate approval mark, and, so far as is reasonably practicable, ensure that every safety cab fitted to a tractor is approved for use with that tractor. No worker shall drive a tractor fitted with a safety cab which he knows is not approved for use with the tractor. He must report to his employer any occasion when the tractor overturns, or when there is damage to the cab or to its fittings, or when there is any defect with the windscreen wiper.

HEALTH AND SAFETY (AGRICULTURE) (POISONOUS SUBSTANCES) REGULATIONS 1975

8.21 These Regulations lay down detailed provisions for safety precautions to be taken when dangerous or poisonous substances are being used for agricultural purposes. It is forbidden to work in certain operations (called the 'Scheduled operations') unless appropriate protective clothing is worn. Also prohibited is the working in a greenhouse or livestock house in which a specified substance is being used unless protective clothing is worn.

8.22 The employer of a worker who is engaged in the scheduled operations shall provide the necessary clothing and maintain it in a good and serviceable condition. Accommodation must be provided for the protective clothing, and for the worker's clothing which is not being worn during working hours, the accommodation being appropriately ventilated so as to ensure that the worker's clothing does not become contaminated. When the scheduled operations are being carried on, the employer must provide adequate and suitable washing facilities, including soap and clean towels (marked 'personal washing only' – the washing of clothing is prohibited) wholesome drinking water with clean vessels, and suitable facilities for ensuring that the worker's food or drink does not become contaminated. At the end of the day's operations, all protective clothing (other than overalls, hoods, respirators or dust masks) must be thoroughly washed, rubber gloves must be washed inside and outside, respirators and dust masks must be cleaned and ventilated. All apparatus used for spraying or applying the specified substance must be kept free from contamination so far as is practicable to do so, tanks and containers must be kept covered, and overalls and hoods which have been worn must be thoroughly washed at least once every six days and, in addition, whenever there is a stain present which may lead to reasonable grounds to fear that a worker may be in danger of poisoning.

8.23 Workers shall not blow, suck, or apply the mouth to any jet, nozzle, sprinkler, etc, which has contained the specified substance, must not use any container marked 'personal washing only' for the washing of any clothing. They must not misuse drinking vessels, eat, drink or smoke unless they have removed all protective clothing, have washed their hands and faces, and are outside the area where the specified substance has been used. They must deposit personal clothing in the accommodation provided, and, at the end of the day's operations, they must remove protective clothing, and wash their hands, faces and necks. Apparatus which has been used and which requires repair must first be washed.

8.24 The employer must keep a register, containing the following particulars:

(a) the names and addresses of every worker employed by him who carries out the scheduled operations;
(b) the number of hours worked on the scheduled operations each day;
(c) the specified substance used;
(d) details of any poisoning suffered by a worker. Further, if a worker is off work for more than three days, and within the preceding fourteen days he had worked for more than sixty hours with certain chemicals, or with any of the specified substances within the previous twenty-eight days, then, unless the employer knows that the absence is not due to any poisoning caused by any of these substances, the employer must notify the inspector forthwith;
(e) details of any condition which led to the granting of a certificate of exemption.

8.25 A copy of the relevant entry in the register shall be given to every worker who leaves the employment in so far as it relates to the last six months of the employment, and this shall be handed to the next employer.

8.26 There are restrictions on the number of hours which can be worked on scheduled operations, provisions relating to training and supervision, and restrictions on the employment of persons below the age of eighteen.

AGRICULTURE (POISONOUS SUBSTANCES) ACT 1952

8.27 Section 6 empowers a HSE inspector to take for analysis any substance which he finds on agricultural premises, and, if practicable, give notice of his intention to do so to the employer of any person working there. He shall divide the sample into three parts, and

(a) if required to do so by the employer, give one part to him;
(b) retain one part for future comparison;
(c) submit one part to an approved analyst.

8.28 If it is not practicable for him to inform the employer of his intention, and he proposes to have the sample analysed, then he will forward one part of the sample to the employer (if he can ascertain his name and address) by registered post or otherwise, together with a notice informing him of the intention to have the sample analysed.

B Construction industry

8.29 Section 127 of the Factories Act applies many of the provisions of that Act to building operations and works of engineering construction undertaken by way of trade or business, or for the purpose of any industrial or commercial undertaking and to any line or siding used in connection therewith which is not part of a railway. The relevant provisions are:

(a) Part I with respect to sanitary conveniences (section 7);
(b) the power contained in sections 50 and 51 so far as they enable the Minister to make Regulations relating to fire prevention and fire fighting;
(c) Part II with respect to steam boilers and air receivers (sections 31–34, 37, 38);
(d) section 82;
(e) Part X with respect to the abstract of the Act, notices, special regulations, general registers and the preservation of records and registers (sections 138, 140, 141);
(f) Part XI relating to certain duties of local authorities (section 153);
(g) Part XII relating to offences (sections 155–171);
(h) Part XIII (sections 172, 174);
(i) Part XIV (sections 175, 176).

8.30 However, the main source of legal rules relating to health and safety in the industry is the four sets of Regulations which are in force. These are:

(a) Construction (General Provisions) Regulations 1961;
(b) Construction (Lifting Operations) Regulations 1961;

(c) Construction (Working Places) Regulations 1966;
(d) Construction (Health and Welfare) Regulations 1966.

These Regulations apply to those premises which come within the scope of section 127 of the Factories Act (above).

Obligations

8.31 The first three sets of Regulations mentioned (i e General Provision, Lifting Operations and Working Places) impose three general duties on every contractor and every employer of workmen. These are:

8.32 (a) To comply with those Regulations as affect him or any workman employed by him. The 'workmen' referred to are not only those who are engaged on the building work, but others employed, e g a nightwatchman (*Field v Perry (Ealing) Ltd*). A main contractor owes the duty towards workmen who are employed by him, but not to those employed by a subcontractor (*Claydon v Lindsay Parkinson Ltd*), nor to the workmen employed by the building owner (*Wingrove v Prestige & Co Ltd*), nor to an independent contractor (*Herbert v Harold Shaw Ltd*). However, the insertion in the Regulations in 1961 of the words 'him or' meant that an independent subcontractor owes a duty to ensure his own safety (*Smith v George Wimpey Ltd*), and that if he failed to do so he could be prosecuted for an offence (obviously, if he injured himself through a breach of his own duty, he could not sue himself for damages!). Although the main contractor thus owes no statutory duty to the employees of the subcontractor, this in no way weakens the obligations owed by the main contractor at common law to ensure that the work is organised in a safe way (*McArdle v Andmac Roofing Ltd*).

8.33 The contractor's obligations do not apply to workmen if their presence in any place is not in the course of performing any work on behalf of the employer and is not expressly or impliedly authorised or permitted by the employer.

8.34 (b) To comply with such requirements of the Regulations as relate to any act or operation performed by any such contractor or employer of workmen. Work is performed by the main contractor if it is being done by the subcontractor as long as the main contractor has not divest himself of control over the execution of the work, but if he has so divest himself, he is not performing the work within the meaning of the Regulations (*Donaghey v Boulton Paul Ltd*).

8.35 (c) Every contractor and every employer of workmen who erects, installs works or uses any plant or equipment, or who erects or alters any scaffolding, shall do so in a manner which complies with the Regulations.

8.36 Additionally, the Regulations impose a duty on every person employed to comply with the legal requirements as relates to the performance of or to the refraining of an act by him, to co-operate in carrying out these Regulations, and if he discovers any defect in any plant or equipment, to report such defect without unreasonable delay to his employer or foreman, or to a person appointed as a safety supervisor.

Construction (General Provisions) Regulations

8.37 The following is an outline of the main provisions of these Regulations.

SAFETY SUPERVISOR

8.38 Every contractor and every employer of workmen who undertakes building or construction work and who normally employs more than twenty persons thereon at any one time (whether or not all those persons are employed on the same site or are all at work at the same time) shall appoint in writing a safety supervisor, who shall be experienced in such operations or works and suitably qualified for the purpose. He shall advise the contractor or employer as to the observance of the requirements for the safety or protection of persons employed which may be imposed by the Factories Act, and exercise a general supervision of the observance of the statutory requirements and of promoting the safe conduct of the work generally. The name of the safety supervisor shall be entered on the Abstract of the Regulations or Factories Act which must be displayed. There is no requirement that the safety supervisor shall be a full-time appointment, but any other duties assigned to him shall not be such as to prevent him from discharging with reasonable efficiency the duties assigned to him under those Regulations. The same person may be appointed for a group of sites, and two or more contractors or employers may jointly appoint the same person to be responsible for a particular site.

SUPPLY AND USE OF TIMBER

8.39 An adequate supply of timber of suitable quality (or other suitable support) shall, where necessary, be provided and used to prevent, so far as is reasonably practicable, danger to any person employed from a fall or dislodgment of earth, rock or any other material forming a side or the roof of or adjacent to any excavation, shaft, earthwork or tunnel.

INSPECTIONS AND EXAMINATIONS OF EXCAVATIONS

8.40 Every part of any excavation, shaft, earthwork or tunnel where persons are employed shall be inspected by a competent person at least once every day during which persons are employed therein, and tunnels and trenches more than 6 foot 6 inches deep shall be inspected at the commencement of each shift.

SUPERVISION OF TIMBERING

8.41 No timbering or other support for any part of an excavation, shaft, earthwork or tunnel shall be erected, substantially added to, altered or dismantled except under the direction of a competent person, and, so far as possible, by competent workmen possessing adequate experience of such work.

MEANS OF EXIT IN CASE OF FLOODING

8.42 If there is any reason to apprehend danger to persons employed therein from rising water or from an eruption of water, there shall be provided, so far as is reasonably practicable, means to enable such persons to reach positions of safety. If the work is likely to reduce the stability or security of any part of the structure (whether temporary or permanent) work shall not be commenced or continued unless adequate steps are taken before or during the progress of the work to prevent danger to any person employed from collapse of the structure or a fall of part of it. Materials shall not be placed or stacked near the edge of an excavation etc so as to endanger persons employed below. Cofferdams and cassions must be of good construction, of suitable and sound material free from patent defect, of adequate strength and properly maintained. They shall not be placed into position, or added to, or altered or dismantled excepf under the immediate supervision of a competent person. No person shall be employed in them unless they have been inspected at least once on the same or preceding day.

EXPLOSIVES

8.43 Explosives shall not be handled or used except by or under the immediate control of a competent person with adequate knowledge of the dangers. Steps must be taken to ensure that when a charge is fired, persons employed are not exposed to the risk of injury from the explosion or from flying materials.

DANGEROUS OR UNHEALTHY ATMOSPHERES

8.44 Where there is any grinding, spraying, cleaning or manipulation of materials which give off any dust or fumes, there must be adequate ventilation or suitable respirators provided. Ventilation must also be secured and maintained in any excavation, pit, hole, adit, tunnel, shaft, cassion or other enclosed or confined space.

WORK ADJACENT TO WATER

8.45 If a person is being carried to or from work by water, proper measures shall be taken to provide for his safe transport. If work is being carried on adjacent to water, and there is a risk of drowning if an employee falls in, suitable rescue equipment shall be provided.

TRANSPORT

8.46 Rails and tracks on which locomotives, trucks or wagons have to move shall have an even running surface, securely fastened, supported so as to prevent undue movement, laid in straight lines or curves on which they can travel without danger of derailment, and provided with a stop or buffer at each end of the track. Locomotives, trucks and wagons shall be of good construction, sound material, adequate strength, free from patent defect and properly maintained. There shall be adequate clearance so that persons are not liable to be trapped or crushed by passing locomotives etc, and suitable recesses where appropriate. If there is no such adequate clearance, effective warnings must be given of approaching locomotives etc. Gantries must be provided with suitable and adequate footways, locomotives must have effective brakes, and suitable arrangements made for derailments. Locomotives shall be fitted with a whistle or other warning device, and only a competent and trained person over the age of eighteen may drive or operate them. Power driven capstans or haulage winches shall have sufficient space for safe working, and sound or visual signals given by the operator to any person who may be endangered by their operation. Mechanically propelled vehicles used for the conveyance of workmen, goods or materials shall be kept in an efficient state, efficient working order and good repair, and may not be used in an improper manner, or loaded so as to interfere with the safe driving. No person shall be permitted to ride in an insecure position, or remain on such vehicle during the loading of loose materials by means of a grab excavator or similar appliance, and when the vehicle is being used for tipping materials, measures shall be taken to prevent it overrunning the edge of any excavation, pit, embankment or earthwork.

DEMOLITION WORK

8.47 Every contractor undertaking demolition operation shall appoint a competent person to supervise the work, and before the work is commenced and during the work all practicable steps shall be taken to prevent danger to persons employed from risk of fire, explosion or leakage or accumulation of gas or vapour, and from the risk of flooding. No part of a building or structure shall be so overloaded with debris or materials as to render it unsafe, and certain types of work may only be done under the immediate supervision of a competent foreman or chargehand with adequate experience of the particular kind of work. All practicable precautions must be taken to avoid danger from collapse of buildings or structures, and adequate shoring shall be undertaken.

OTHER MATTERS

8.48 Flywheels, moving parts of prime movers, transmission machinery and every dangerous part of other machinery (whether or not driven by mechanical power) shall be securely fenced, and prime movers and mechanically driven machines constructed after the commencement of the Regulations in 1962 must have certain specific parts securely fenced. All practicable steps shall be taken to prevent dangers to persons employed from any live electric cable or apparatus, and measures shall be taken to prevent steam, smoke or other vapour generated from the site from obscuring the work, scaffolding, machinery or other plant or equipment where any person is employed. Steps shall be taken to prevent any person who is working from being struck by any falling material or article. Every workplace and its approaches shall be adequately and suitably lit. No timber or materials with projecting nails shall be used, and loose materials shall not be left so as to unduly restrict the passage of persons, but shall be removed and stacked or stored. Materials shall not be insecurely stacked in a place where they may be dangerous. Temporary structures shall be of good construction and adequate strength and stability. All practicable steps shall be taken to prevent danger from the collapse of any building or structure during its temporary weakness. Iron or steelwork which has been painted or cement washed shall not be moved or manipulated unless the paint or cement is dry. Helmets or crowns used for pile-driving shall be of good construction, of sound and suitable material of adequate strength and free from patent defect. No person shall be employed to lift, carry or move a load so heavy as to be likely to cause injury to him. Form 91 (report of examination of excavations and of cofferdam or cassion see above) shall be kept on the site, and all other reports or documents required to be kept shall at reasonable times be open to inspection by HSE inspectors.

Construction (Lifting Operations) Regulations 1961

8.49 The following is an outline of the main provisions of these Regulations.

LIFTING APPLIANCES

8.50 Every lifting appliance and all appliances which support lifting appliances shall be of good mechanical construction, sound material, adequate strength free from patent defect, properly maintained, and as far as the construction permits be inspected at least once in every week by the driver or other competent person. The result shall be recorded on Form 91. If the lifting appliance has a travelling or slewing movement there shall be an unobstructed passageway of not less than 2 feet wide between the moving parts and any guardrails, unless all reasonable steps are taken to prevent the access of any person to such place. Where a platform is provided, it shall be of sufficient area, close planked and provided with a safe means of access, with sides provided with suitable guardrails. The driver of every power driven lifting appliance shall be provided with a suitable cabin which shall afford him adequate protection from the weather and have ready access to parts which need periodic inspection or maintenance, and shall not prevent him from having a clear and unrestricted view. Drums and pulleys shall be of suitable diameter, and every crane, grab and winch shall have efficient brakes or other safety device which will prevent the fall of the load when suspended. A safe means of access to and egress from any place where a person has to work on examination, repair or lubrication shall so far as is reasonably practicable, be provided and maintained. Poles or beams which support pulley blocks and gin wheels shall be of adequate strength and adequately and properly secured. Appropriate precautions shall be taken to ensure the stability of lifting appliances used on a soft or uneven surface or on a slope. Before a crane is erected, the anchorage must be examined by a competent person, and after erection there will be a further test involving a load which is 25 percent above the maximum load to be lifted by the crane. A report of such tests and the results shall be made.

8.51 Rail mounted cranes shall be supported on a firm and even surface jointed by fish plates or double chairs, securely fastened, laid in straight lines or curves which will not cause derailment, and provided with stops or buffers at each end. Bogeys, trolleys or wheeled carriages on which a crane is mounted shall be of good construction, adequate strength, suitable, made from sound material free from patent defect and properly maintained. Cranes with derricking jibs shall have effective interlocking arrangements to prevent the disengagement. A crane shall not be used for any purpose other than the raising or lowering of a load vertically unless no undue stress is imposed and the use is supervised by a competent person. Cranes shall not have any timber structure and shall be erected under the supervision of a competent person.

8.52 Lifting appliances shall not be operated except by a trained and competent person (unless under the direction of a qualified person for the purpose of training) and no person under the age of eighteen (other than as a trainee) shall be employed to operate the appliance or give signals to the operator. If the operator does not have a clear and unrestricted view, one or more competent persons shall be appointed and suitably stationed to give the necessary signals. Every signal given shall be distinctive in character, and if devices are used for giving sound, colour or light signals, these shall be properly maintained. Cranes, crabs and winches shall not be used unless tested by a competent person within the previous four years, and a further test shall take place if they have been subject to any substantial alteration or repair affecting its strength or stability.

8.53 No crane, crab, winch, pulley block or gin wheel shall be used unless a test certificate has been obtained stating (a) the safe working load, (b) the safe radii or maximum radii. Jib cranes shall have an automatic safe load indicator which has been tested before the crane is taken into use, and inspected once each week it is being used. Lifting appliances generally shall not be loaded beyond the safe working load, and where more than one lifting appliance is required to raise a load, the equipment shall be so arranged that no appliance shall be loaded beyond it safe working load or be rendered unstable. Such operations must be conducted under the supervision of a competent person.

8.54 Chains, ropes or lifting gear must be of good construction, sound materials, adequate strength, suitable quality, free from patent defect, tested and examined before use, and marked with its safe working load and means of identification. There are further provisions designed to ensure safety when using chains, rings, hooks etc. Part V of the Regulations deals with hoistways, platforms and cages, which must have substantial enclosures, gates, safety devices, and safe working loads. Hoists must be thoroughly examined by a competent person every six months, and a report sent to the district inspector within twenty-eight days where the test or examination reveals that certain repairs are required.

Construction (Working Places) Regulations 1966

8.55 The following is an outline of the main provisions of these Regulations.

8.56 So far as is reasonably practicable, there shall be a suitable and sufficiently safe means of access to and egress from every place at which any person at any time works and every place where a person works shall also be kept safe. No scaffold shall be erected or substantially added to or altered or be dismantled unless under the immediate supervision of a competent person with adequate experience of such work. Materials used for scaffolding shall be inspected each time they are used. Every scaffold shall be of good construction, of suitable and sound material and of adequate strength. Sufficient material shall be provided, timber which is used shall not be painted so that defects cannot be seen, and metal parts must be free from corrosion. Defective materials shall not be used.

8.57 There are a number of provisions relating to the erection and dismantling of scaffolding, the materials which may or may not be used, supervision by competent persons, maintenance, stability, slung scaffolds, cantilever jibs and suspended scaffolds, cages, skips, trestle scaffolds, etc. There are also detailed requirements relating to working platforms, gangways and runs, with provisions for guard rails and toeboards. If any person is liable to fall a distance of more than 6 foot 6 inches, guardrails must be provided. Ladders must be of good construction, of suitable and sound material and of adequate strength, securely fixed near to its upper resting place, or, if that is impracticable, at or near the lower end. Work on sloping roofs may only be carried out by suitable workmen, using crawling boards or crawling ladders. In certain circumstances, safety nets or safety sheets must be provided.

Construction (Health and Welfare) Regulations 1966

OBLIGATIONS

8.58 Except as otherwise provided, it is the duty of every contractor to comply with such requirements as affect any person employed by him (compare this with the obligations imposed by the other three sets of Regulations, above). If the contractor makes effective arrangements with another contractor on the site, or similar arrangements with another person for enabling persons employed to have adequate access to and use of facilities and which are reasonably accessible, then the contractor will have fulfilled his obligations in respect of those stated matters.

8.59 The following is an outline of the requirements of the Regulations.

FIRST AID

8.60 The Regulations dealing with first aid have been revoked and replaced by the Health and Safety (First-Aid) Regulations 1981 (see chapter 6).

SHELTERS ETC

8.61 At or in the immediate vicinity of every site there shall be conveniently accessible for the use of persons employed adequate and suitable accommodation for taking shelter during interruptions of work owing to bad weather, and for depositing clothing not worn during working hours. If there are more than five persons employed on the site, there must be adequate and suitable means for enabling persons employed to warm themselves and to dry their clothing, and if there are five or less so employed, such arrangements must be made as are reasonably practicable for these purposes. Similar accommodation and drying facilities must be provided for protective clothing used for work. Dry facilities for taking meals and for boiling water must be provided and, if more than ten persons are employed and heated food is not available on site, adequate facilities for heating food. Wholesome drinking water must also be provided. All accommodation must be kept in a clean and orderly condition and not used for the storage of materials or plant.

WASHING FACILITIES

8.62 If an employee works on site for more than four consecutive hours, there must be adequate and suitable washing facilities, and if more than twenty employees are likely to be working on site for more than six weeks there shall be provided adequate troughs, basins or buckets, adequate means of cleaning and drying and a sufficient supply of hot and cold or warm water. If there are more than 100 likely to be employed on site for more than twelve months a minimum scale for these facilities is laid down. Suitable sanitary conveniences which are conveniently accessible shall be provided if there are more than twenty-five persons employed on site, which shall be ventilated, under cover, partitioned off so as to ensure privacy, and screened off with a proper door and fastening. Adequate and suitable protective clothing shall be provided for any person who by the reason of the nature of the work is required to continue working in the open air during rain, snow, sleet or hail. All the above facilities shall so far as is reasonably practicable have a safe means of access or egress, and every such place shall equally be made and kept safe.

8.63 The following is a summary of the examinations and inspections which are required under the Construction Regulations generally.

	Nature of examination or inspection	*When undertaken*	*By whom*
(1)	Every part of any excavation, shaft, earthwork or tunnel	at least once on every day during which persons are employed therein	competent person
(2)	Face of every tunnel and the working end of every trench more than 6 ft 6 in deep, and the base or crown of every shaft	commencement of every shift	competent person
(3)	Thorough examination of parts of any excavation, shaft, earthwork or tunnel in the region of a blast	(a) after explosives have been used nearby in a manner likely to affect the strength or stability of timbering or other support; (b) since timbering or other support has been substantially damaged; (c) in the region of any unexpected fall of rock, earth or other material;	competent person

	Nature of examination or inspection	*When undertaken*	*By whom*
		(d) every part within the immediately preceding 7 days	
(4)	Timbering or support material used for excavation, shaft, earthwork or tunnels	each occasion before use	competent person
(5)	Materials used for the construction of a cofferdam or cassion	each occasion before use	competent person
(6)	Cofferdams or cassions must be inspected	(a) on the day or day before a person is employed therein; (b) since explosives have been used in a manner likely to affect the strength or stability; (c) since the cofferdam or cassion has been substantially damaged; (d) in the 7 days immediately preceding persons working therein	competent person
(7)	Suitable testing of the atmosphere in an excavation, pit, hole, adit, tunnel shaft, cassion or other enclosed or confined space	where there is reason to apprehend that the atmosphere is poisonous or asphyxiating	by or under the immediate supervision of a competent person
(8)	Periodic examination of hoists forming part of permanent equipment	before use for carrying persons	competent firm of lift engineers
(9)	Every part of lifting appliances, all working gear, plant or equipment used for anchoring or fixing	as far as the construction permits inspection	by the driver (if competent) otherwise by a competent person
(10)	The whole of the appliances for the anchorage or ballasting of a crane	each occasion before the crane is erected	competent person

	Nature of examination or inspection	*When undertaken*	*By whom*
(11)	Securing of the anchorage or adequacy of the ballasting of cranes	before use, after each erection, removal or adjustment	competent person
(12)	Anchorage arrangements and ballast of cranes	after exposure to weather conditions likely to affect stability	competent person
(13)	Thorough examination of crane, grab or winch	within four years prior to use	competent person
(14)	Thorough examination of pully block, gin wheel, or sheer legs	if used to raise or lower a load weighing more than 1 ton	competent person
(15)	Thorough examination of crane, grab or winch, pulley block, gin wheel, sheer legs	after any substantial alteration	competent person
(16)	Thorough examination of lifting appliances	within previous 14 months, or since undergone any substantial alteration or repair	competent person
(17)	Approved type of automatic safe load indicators on jib cranes	tested after erection or installation and before use	competent person (other than the crane driver)
(18)	Approved type of automatic safe load indicators on jib cranes	inspected once a week when crane is in use	driver (if competent) or competent person
(19)	Approved type of automatic safe load indicators on mobile cranes	tested before use	competent person
(20)	Approved type of automatic safe load indicators on jib cranes	inspected once a week when crane is in use	competent person
(21)	Chains, ropes, lifting gear used in raising or lowering or as a means of suspension	tested and examined before use	competent person

	Nature of examination or inspection	*When undertaken*	*By whom*
(22)	Chains, ring, link, hooks, plate clamps, shackle, swivel or eye-bolt which has been lengthened, altered, or repaired by welding	tested and thoroughly examined before use	competent person
(23)	Chains, ropes and lifting gear in regular use	thorough examination at least once in 6 months	competent person
(24)	Thorough examination of hoists which have been manufactured or substantially altered or repaired after March 1962	before being used	competent person
(25)	Thorough examination of hoists used for carrying persons	since erected or height altered	competent person
(26)	Thorough examination of hoists	at least once each 6 months	competent person
(27)	All material used for any scaffold	before taken into use	competent person
(28)	Scaffolds (including boatswain's chair, cage, skip or similar plant or equipment)	inspected within 7 days prior to use	competent person

9 Compensation for injuries at work

9.1 Inevitably, an employee who is injured in the course of his employment will seek some form of compensation from the person (if any) who was responsible for those injuries, or look to the state to provide some assistance. If an employee is killed, his dependants will also be seeking some form of financial recompense for themselves. It is this area of the law which in the past has tended to dominate all other considerations of health and safety, as compensation, rather than prevention, became the main function of the law. This attitude prompted the Robens Committee to attempt some shift in the impact of the legislation.

9.2 It is a truism to say that legal decisions reflect the current social and economic forces of the day, and this explains some of the changes in judicial attitudes which have taken place from time to time. Further, the impact of state and private insurance schemes (especially Employers' Liability Insurance) has cast a powerful shadow over strict legal reasoning, for there is a natural tendency to seek some way in which the injured employee can be assisted financially, and, after all, if an insurance company had to foot the bill, this could easily be recouped by a minute increase in general premiums! This benevolent attitude even reached employers, and there are many cases where liability is admitted or where valiant attempts were made to admit their own negligence so as to enable an injured employee to obtain compensation (e g see *Hilton v Thomas Burton (Rhodes) Ltd*). But there were limits, too, on judicial credulity, which compelled some judges to hold that they could no longer equate the relationship between employer and employee with that of nurse and imbecile child (Lord Simmonds in *Smith v Austin Lifts Ltd*) or that of schoolmaster and pupil (Devlin LJ in *Withers v Perry Chain Ltd*).

9.3 The basis of the employer's duty towards his employees stems from the existence of a contract of employment. It is an implied term of that contract that the employer will take reasonable care to ensure the safety of his employees (*Matthews v Kuwait Bechtel Corporation*), and an employer who fails to fulfil that duty is in breach of that contract (*British Aircraft Corporation v Austin*). However, from the point of view of an injured employee there is little advantage in suing in contract. Practically all modern cases are brought under the law of tort, in particular for the tort of negligence which, since the famous case of *Donoghue v Stevenson* in 1932 consists of three general ingredients, namely (a) there is a general duty to take care not to injure someone whom one might reasonably foresee would be injured by acts or omissions, (b) that duty is broken if a person acts in a negligent manner, and (c) the breach of the duty must cause injury or damage. The existence of a duty-situation between employer and employee has been long recognised, and most of the cases turn on the second point, i e was the employer in fact negligent?

9.4 The liability of the employer may come about in two ways. First, he will be responsible for his own acts of negligence. These may be his personal failures or (since many employers nowadays are artificial legal entities, i e limited companies or other types of corporations) due to the wrongdoings of the various acts of management acting as the *alter ego* of the employer. Second, the employer may be liable vicariously for the wrongful acts of his employees which are committed in the scope of their employment and which cause injury or damage to others.

9.5 Compensation may be obtained under one or more of three headings. The first is for the injured employee (or his personal representatives, if he has died) to bring an action at common law. This will be for either (a) a breach by the employer of a duty laid down by statute, or (b) a breach of the duty owed by the employer at common law to ensure the health and safety of his employees. Frequently, a claim will be presented under both headings simultaneously. With two strings to his bow, it matters not if he wins under either heading, or both. (Of course, if he wins under both, he will only get one lot of damages.) It is regarded as a misfortune if he fails under both headings, for he has then lost his main financial solace. It was the uncertainties of litigation in this area of the law, the difficulties of establishing satisfactory evidential standards in court hearings held years after an incident had occurred, the legal expenses and complexities, and the social injustices caused, which were among the main reasons for the appointment of the Pearson Commission in 1974. Its report, however, did not recommend any major changes in the compensation system.

9.6 The second remedy for an injured employee is the automatic recourse to the National Insurance (Industrial Injuries) scheme operated by the state in one form or another since 1911 (see now the Social Security Act 1975), which provides a system of benefits for a person injured at work or who contracts a prescribed disease. There are also pensions or lump sum gratuities paid in respect of permanent disabilities. This scheme is in additional to the right to sue at common law, although benefits payable will be taken into account when fixing damages. However, the state scheme operates as of right, irrespective of the existence of fault or blame on the part of any person, including the injured worker.

9.7 Third, in appropriate circumstances, there are other schemes which may be resorted to, such as those operated by the Criminal Injuries Compensation Board, the Motorists Insurance Bureau or (rarely used), the power of the courts to award compensation.

A Claims at common law

1 Breach of statutory duty

9.8 When Parliament lays down a duty for a person to perform, it will usually ensure that there is an appropriate sanction to enforce that duty. This sanction will normally take the form of some sort of punishment in the criminal courts. The further question will thus arise; can a person who has been injured by the failure of another to perform that statutory duty bring an action based on that failure? After some hesitation, British courts upheld an action for the tort of breach of statutory duty, although the full extent of legal liability is not entirely settled. There is no automatic presumption that all breaches of statutory duties are actionable in civil courts; it is necessary to examine the purposes and objects of the statute in question, seek the intentions of Parliament (as discovered from the words used in the Act, not the statements of Ministers or MPs in Parliamentary debates etc), and ascertain the class of persons for whose benefit the Act was passed. If the injured party has suffered the type of harm the Act was designed to eliminate, it would not be unreasonable to grant him a remedy in respect of a breach of the statutory duty. The first case in which these propositions were accepted was *Groves v Lord Wimborne*, where a statute provided that an occupier of a factory who did not fence dangerous machinery was liable to a fine of up to £100. A boy employed in the factory was caught in an unfenced cog wheel, and his arm was amputated. It was held that the criminal penalty was irrelevant to civil liability, and the claim for a breach of statutory duty succeeded.

9.9 However, not every breach of statutory duty is actionable (*Cutler v Wandsworth Stadium Ltd*), and though the point appears to be well settled so far as health and safety legislation is concerned, there are one or two areas where there may still be an element of doubt. The real difficulty is encountered when dealing with some of the welfare provisions in the Factories Act, OSRPA etc. For example, in *Ebbs v James Whitson & Co Ltd* the question arose as to whether the plaintiff could sue in respect of a breach of section 4(1) of the Factories Act (securing adequate ventilation, see chapter 4) and Denning LJ (as he then was) expressly reserved his opinion on this point. However, similar claims have succeeded in a number of cases since then, though the point was either conceded or not argued. For example, in *Nicholson v Atlas Steel Ltd* a worker contracted pneumoconiosis as a result of exposure to a silica dust, from which he subsequently died. It was held that his widow was entitled to recover damages based on a breach of section 4 of the Factories Act in respect of a failure to provide adequate ventilation. Civil remedies have also been granted in other cases dealing with Part I of the Factories Act (e g *Lane v Gloucestershire Engineering Co Ltd*, a case under section 5) and it would appear that the doubts of Lord Denning are misplaced. However, the point is less clear when considering some of the welfare provisions of Part III of the Act (see chapter 4), although again, some actions have succeeded without argument on this point (e g *McCarthy v Daily Mirror*: section 59 – suitable accommodation or clothing). Similar considerations will presumably apply to other health and safety legislation, e g OSRPA, but there does not appear to be any direct decision on the point. It would seem that the accepted view is that welfare provisions are of two distinct types; first, those which are concerned with health and safety, and those which are dealing with welfare and comfort. It may well be that no civil action will lie in respect of a breach of the latter kind of statutory provisions. Thus, if an employer fails to provide adequate washing facilities (Factories Act, section 58) this duty cannot be enforced *per se* by an employee, only by the Inspectorate. If, as a result of the failure by the employer to provide washing facilities an employee contracts dermatitis, he may be able to recover damages

(*Reid v Westfield Paper Co Ltd*). For a contrary viewpoint see *Clifford v Challen & Son Ltd*.

9.10 It will be recalled that section 47 of HSWA provides that the no civil action may be brought in respect of a breach of the general duties and obligations contained in sections 2–8, but that a breach of health and safety Regulations made under the Act shall be actionable except in so far as the Regulations provide otherwise.

ELEMENTS OF THE TORT OF BREACH OF STATUTORY DUTY

9.11 The first requirement is that the plaintiff must show that he is within the class of persons for whose benefit the duty was imposed. This will depend entirely on the provision in question. Thus there are provisions in the Factories Act which are designed to protect all persons who work in a factory, whether or not they are the employees of the occupier, and whether or not they are doing the employer's work or their own (*Uddin v Associated Portland Cement Ltd*). Other provisions may be more limited in their scope. Thus in *Hartly v Mayoh & Co* a fireman was electrocuted whilst fighting a fire at the defendant's premises. The widow sued in respect of a breach by the defendant of a statutory Regulation. Her claim failed. The provisions in question were designed to protect 'persons employed' in the premises, and the fireman was not within this class of person. In *Reid v Galbraith's Stores* it was held that the provisions of OSRPA did not apply to a customer who was visiting the shop, as the Act was concerned with people who work on the premises.

9.12 In so far as Regulations are concerned, the right to bring a civil claim in respect of a breach is probably even more circumscribed, for sometimes these will be designed to protect a person who is performing a particular type of work, and a person who is not engaged on that process, but who may be injured as a result of a breach of the Regulation, will not generally be entitled to compensation under this heading. The reason is that in the past it has been presumed that Parliament only intended that the power to make Regulations shall be exercised within particular limits, and the courts will not therefore go outside those limits. However, the power to make Regulations under section 15 of HSWA is not so restricted and thus they may include a wider category of affected persons, as appropriate. It is clear that when Regulations are made for the benefit of a particular group of persons, only those who are in that group can take advantage of the statutory protections (*Canadian Pacific Steamships Ltd v Bryers*).

9.13 Second, the injury must be of a kind which the statute was designed to prevent. In *Close v Steel Co of Wales* a workman was injured by a part of dangerous machinery which flew out of a machine. His claim, based on a breach of section 14 of the Factories Act (duty to fence dangerous parts of machinery, see chapter 4) failed. The object of section 14 is to prevent the worker from coming into contact with the machine, not to stop the parts of the machine from coming into contact with the worker. The purpose of fencing is not to keep the machine or its products inside the fence!

9.14 The third requirement of the tort is that the defendant must be in breach of that duty. This involves a consideration of the duty imposed, the person upon whom it is placed, and the steps taken to perform that duty. Thus in *Chipchase v British Titan Products Ltd*, Regulations provided that every working platform from which a person is liable to fall more than 6 feet 6 inches shall be at least 34 feet wide. A worker fell from a platform which was only 9 inches wide, but which was 6 feet from the ground. The plaintiff's case obviously failed, as there was no breach of duty by the defendant. If the statute or Regulations imposes absolute duties, then these must be observed irrespective of the inconvenience caused (*John Summers v Frost*), but if these are qualified, e g 'so far as is reasonably practicable' etc, it is a question of fact in each case as to whether or not the defendant has complied with that standard.

9.15 Finally, it must be shown that the breach of the duty caused the damage. This is the causation rule, which may be illustrated by the decision in *McWilliams v Sir William Arrol & Co Ltd*. Here, the employers provided safety belts for steel erectors on a site. As the belts were not being used, they were taken to another site. A steel erector on the first site fell from a scaffolding, and was killed. Although the employers were clearly in breach of their statutory duty to provide the safety belts, they were not liable for damages. Even if they had provided the belts, there is nothing to suggest that this workman (who had never used them before) would have worn them on the day he was killed. Thus, the breach of duty did not cause the damage; it would have occurred anyway.

2 Negligence

9.16 At common law, the employer is under a duty to take reasonable care for the health and safety of his employees. This duty is a particular aspect of the general law of negligence, which requires everyone to ensure that his activities do not cause injury or damage to another through some act of carelessness or inadvertence (see *Donoghue v Stevenson*, para 9.3).

THE PERSONAL NATURE OF THE DUTY

9.17 The duty at common law is owed personally by the employer to his employees, and he does not escape that duty by showing that he has delegated the performance to some competent person. In *Wilsons and Clyde Coal Co v English* the employer was compelled by law to employ a colliery agent who was responsible for safety in the mine. Nonetheless, when an accident occurred, the employer was held liable. Thus it can never be a defence for an employer to show that he has assigned the responsibility of securing and maintaining health and safety precautions to a safety officer or other person. He can delegate the performance, but not the responsibility.

9.18 Further, the duty is owed to each employee as an individual, not to employees collectively. Greater precautions must be taken when dealing with young or inexperienced workers, and with new or untrained employees, than one might take with more responsible staff, for the former may require greater attention paid to their working methods, or may need more supervision (*Byers v Head Wrightson & Co Ltd*). In *Paris v Stepney Borough Council* the plaintiff was employed to scrape away rust and other superfluous rubbish which had accumulated underneath buses. It was not customary to provide goggles for this kind of work. However, the plaintiff had only one good eye, and he was totally blinded when a splinter entered his good eye. It was held that the employers were liable for damages. They should have foreseen that there was a risk of greater injury to this employee if he was not given adequate safety precautions, and the fact that they may not have been under a duty to provide goggles to other employees was irrelevant.

9.19 A higher standard of care is also owed to employees whose command of English language is insufficient to understand or comply with safety instructions, to ensure that as a result they do not cause injuries to themselves or to others. In *James v Hepworth and Grandage Ltd* the employers put up large notices urging employees to wear spats for their personal protection. Unknown to them, one of their employees could not read, and when he was injured he claimed damages from his employer. His claim failed. He had observed other workers wearing spats, and his failure to make any enquiries led the court to believe that even if he had been informed about the contents of the notice, he would still not have worn the spats. But with the growth of foreign labour in British factories, the problem is one for obvious concern, especially as immigrants tend to concentrate in those industries which are most likely to have serious safety hazards. The responsibility of the safety officer towards such employees is likely to be a very onerous one (*Hawkins v Ian Ross (Castings) Ltd*).

9.20 The duty is owed by the employer to his employees, but this latter term is not one which is very precise, and a number of problems have arisen. In *Ferguson v John Dawson & Partners Ltd* a man agreed to work on the 'lump' as a self-employed bricklayer, but when he was injured, he sued, claiming the employers' owed a duty to him as an employee. By a majority, the Court of Appeal upheld his claim, holding that it was the substance of the relationship, not the form, which was the determining factor in deciding whether or not a person was an employee in the legal sense.

9.21 As a general rule, each employer must ensure the safety of his own employees, and is not responsible in his capacity of an employer for the safety of employees of other employers. However, where a number of employees from different firms are employed on one job, there is a duty to co-ordinate the work in a safe manner (*McArdle v Andmac Roofing Co*).

9.22 In some circumstances, the employer may be liable for the safety of a loaned employee, i e one who has been seconded to work for him, or, conversely there are circumstances when the employer will still be responsible for the safety of his employee who has been seconded to someone else. The courts will look to see which of the two employers retains the right to control the workman (*Mersey Docks and Harbour Board v Coggins and Griffiths*). Thus if a workman is loaned (together with an expensive piece of equipment, such as a crane) or he is an expert in his job so that the second employer cannot exercise any control over the way the work is done, it will be rare that the courts will infer that the right of control vested in the first employer has been transferred to the second. But if an unskilled workman is loaned, it would be easier to infer that there has also been a transfer of the legal responsibility to the second employer (*Garrard v Southey & Co*). Thus once the right to control has been transferred, the second employer must accept the duty of care to the transferred employee.

THE EXTENT OF THE DUTY

9.23 The standard of care which must be exercised by the employer is 'The care which an ordinary prudent employer would take in all the circumstances' (*Paris v Stepney Borough Council*). The employer does not give an absolute guarantee of health or safety, he only undertakes to take reasonable care, and will be liable if there is some lack of care on his part or in failing to foresee something which was reasonably foreseeable. The employee, for his part, must be prepared to take steps for his own safety and look after himself, and not expect to be able to blame the employer for everything which happens. In *Vinnyey v Star Paper Mills Ltd*, the plaintiff was instructed by the foreman to clear and clean a floor area which had been made slippery by a viscous fluid. The foreman gave him proper equipment and clear instructions. The plaintiff was injured when he slipped on the floor while doing the work, and it was held that the employers were not liable. There was no reasonably foreseeable risk in performing such a simple task, and they had taken all due care. In *Lazarus v Firestone Tyre and Rubber Co Ltd* the plaintiff was knocked down in the general rush to get to the canteen. It was held that this was not the sort of behaviour which could be protected against.

9.24 If an employer does not know of the danger, and could not reasonably be expected to know, in the light of current knowledge available to him, or did not foresee that there was a potential hazard, and could not reasonably be expected to foresee it, he will not be liable. In *Down v Dudley, Coles Long Ltd* an employee was partially deafened by the noise which came from a cartridge assisted hammer gum. At the then state of medical knowledge (i e in 1964) a reasonable employer would not have known of the potential danger in using this particular piece of equipment without providing adequate safety precautions, and the employer was held to be not liable for the injury. Clearly, with the wide dissemination of literature on noise hazards nowadays, a different conclusion would be drawn on these facts, although the law remains the same.

9.25 Once a danger has been perceived, the employer must take all reasonable steps to protect the employees from the consequences of those risks which have hitherto been unforeseeable. In *Wright Rubber Co Ltd, Cassidy v Dunlop Rubber Co Ltd* the employers used an anti-oxidant known as Nonox S from 1940 onwards. The manufacturers then discovered that the substance was capable of causing bladder cancer, and informed the defendants that all employees should be screened and tested. This was not done for some time, and thus the employers, as well as the manufacturers were held to be liable to the plaintiffs.

9.26 The matter was summerised by Swanwick J in *Stokes v GKN Ltd:*

(a) the employer must take positive steps to ensure the safety of his employees in the light of the knowledge which he has or ought to have;
(b) the employer is entitled to follow current recognised practice unless in the light of common sense or new knowledge this is clearly unsound;
(c) where there is developing knowledge, the employer must keep reasonably abreast with it, and not be too slow in applying it;
(d) if he has greater than average knowledge of the risk, he must take more than average precautions;
(e) he must weigh up the risk (in terms of the likelihood of the injury and possible consequences) against the effectiveness of the precautions to be taken to meet the risk, and the cost and inconvenience.

9.27 Applying these tests, if the employer falls below the standards of a reasonable and prudent employer, he will be liable.

The threefold nature of the duty

9.28 Recent cases have stressed that there is only one single duty on the part of the employer, namely to take reasonable care. However, we may conveniently analyse that duty under three headings.

i *Safe plant, appliances, and premises*

9.29 All tools, equipment, machinery, plant which the employee uses or comes into contact with, and all the employer's premises shall be reasonably safe for work. Thus a failure to provide the necessary equipment (*Williams v Birmingham Battery and Metal Co*) or providing insufficient equipment (*Machray v Stewarts and Lloyds Ltd*) or providing defective equipment (*Bowater v Rowley Regis Corporation*) will amount to a breach of the duty. There must be a proper and adequate system of inspection and testing, so that defects can be discovered and reported (*Barkway v South Wales Transport Co*), and then remedied (*Monaghan v Rhodes &*

Son). In *Bradford v Robinson Rentals*, a driver was required go on a 400 mile journey during a bitterly cold spell of weather in a van which was unheated and had cracked windows. He suffered frost-bite, and his employers were held liable for failing to provide suitable plant. Before putting secondhand machinery into use, it should be checked to make sure that it is serviceable (*Pearce v Round Oak Steel Works Ltd*). If unfenced machinery is liable to eject parts of the machine or materials used by the machine, then a failure to erect suitable and effective guards may well constitute negligence at common law (*Close v Steel Co of Wales*) irrespective of any liability for a breach of section 14 of the Factories Act (*Kilgollan v Cooke & Co Ltd*). If the equipment is inherently dangerous, extra precautions must be taken (*Naismith v London Film Productions Ltd*).

9.30 However, if an employer purchases tools or equipment from a reputable supplier, and has no knowledge of any defect in them, he will have performed his duty to take care, and cannot be held liable for negligence (see *Davie v New Merton Board Mills*). An employee who was injured in consequence could only pursue his remedy against the person responsible for the defect under the general law of negligence (*Donoghue v Stevenson*). In practice, this would frequently be difficult or impossible. The employee would not have the time or resources to do this; it may be that the negligence was due to the acts of a foreign manufacturer, or stevedores at the docks etc. In view of these problems, the law was changed with the passing of the Employers Liability (Defective Equipment) Act 1969. This provides that if an employee suffers a personal injury in the course of his employment in consequence of a defect in equipment provided by his employer for the purposes of his employer's business, and the defect is attributable wholly or partly to the fault of a third party (whether identified or not), then the defect will be deemed to be attributable to the negligence of the employer. Thus, in such circumstances, the injured employee would sue the employer for his 'deemed' negligence, and the latter, for his part, would attempt to recover the amount of damages he has paid out from the third party whose fault it really was. Since the insurance company is the real interested party in such matters, it is they, rather than the employer, who will attempt to make such recovery. The Employers Liability (Compulsory Insurance) Act was also passed in 1969 to ensure that all employers have valid insurance cover to meet personal injuries claims from their employees, and a certificate to this effect must be displayed at the employers' premises (see chapter 6).

9.31 If an employer is aware of any defect in tools etc which have been purchased from outside he should withdraw them from circulation. In *Taylor v Rover Car Co Ltd* a batch of chisels had been badly hardened by the manufacturers. One had, in fact, shattered without causing an injury, but the rest of the batch were still being used and another chisel shattered, injuring the plaintiff in his eye. The employers were held liable.

9.32 The employer must also ensure that the premises are reasonably safe for all persons who come on to the premises (under the Occupiers Liability Act 1957) as well as for his employees in particular. In *Paine v Colne Valley Electricity Supply Co Ltd* an employee was electrocuted because a kiosk had not been properly insulated, and the employers were held liable. However, it must be stressed that the employer need only take reasonable care, and this is a question of fact and degree in each case. In *Latimer v AEC Ltd* a factory floor was flooded after a heavy storm, and a mixture of oil and water made the floor slippery. The employers put down sand and sawdust, but there was not enough to treat the whole of the factory in this way, and the plaintiff was injured. The employers were held not liable. The danger was not grave enough to warrant closing down the whole factory (which would have been unreasonable, bearing in mind that the risk was fairly minimal). However, had there been dangers because the structure had been damaged, different considerations would have applied.

9.33 The employer cannot be responsible for the premises of other persons where his employees have to work, but as he still owes to them a duty of care, he must ensure that a safe system of work is laid down.

ii *Safe system of work*

9.34 The employer is responsible for the overall planning of the work operations so that it can be carried out safely. This includes the layout of the work, the systems laid down, training and supervision, the provision of warnings, protective clothing, protective equipment, special instructions, etc. Regard must be had for the fact that the employee will be forgetful, careless, as well as inadvertent, but the employer cannot guard against outright stupidity or perversity.

9.35 Examples of a failure to provide a safe system of work abound. In *Barcock v Brighton Corporation*, the plaintiff was employed at an electrical sub-station. A certain method of testing was in operation which was unsafe, and in consequence the employee was injured. The employers were held liable. If there are safety precautions laid down, the employee must know about them; if safety equipment is provided, it must be available for use. In *Finch v Telegraph Construction and Maintenance Co Ltd* the plaintiff was employed as a grinder. Goggles were provided for this work, but no-one told him where to find them. The employers were held liable when he was injured by a piece of flying metal. Whether a system is safe in any particular case will be a question of fact, to be decided on the evidence available.

9.36 The more dangerous the process, the greater is the need to ensure that it is safe. On the other hand, the employer cannot be expected to take over-elaborate precautions when dealing with simple and obvious dangers (*Vinnyey v Star Paper Mills*). A situation which gives rise to some legal difficulties is where the employer provides safety precautions or equipment, but the employee fails or refuses to use or wear them. Is the duty a mere passive one, to provide, and do no more? Or is it a more active one, to exhort, progandandise, or even compel their use? It is suggested that the answer to these questions can be summarised in four propositions.

9.37 (1) If the risk is an obvious one, and the injury which may result from the failure to use the precautions is not likely to be serious, then the employer's duty is a mere passive one of providing the necessary precautions, informing the employees of their presence, and leaving it to them to decide whether or not to use them. In *Qualcast (Wolverhampton) Ltd v Haynes* an experienced workman was splashed by molten metal on his legs. Spats were available, but the employers did nothing to ensure they were worn. The injury, though doubtless painful, was not of a serious nature, and the employers were held not liable.

9.38 (2) If the risk is that of a serious injury, then the duty of the employer is a higher one of doing all he can to ensure that the precautions are used. In *Nolan v Dental Manufacturing Co Ltd* a toolsetter was injured when a chip flew off a grinding wheel. Because of the seriousness of the injury should one occur, it was held that the employers should have insisted that protective goggles should be worn.

9.39 (3) If the risk is an insidious one, or one the seriousness of which the employee would not readily appreciate, then again, it is the duty of the employer to do all he can by way of propaganda, constant reminders, exhortation, education etc to try to get the employees to use the precautions. In *Berry v Stone Manganese* the plaintiff was working in an environment where the noise levels were dangerously high. Ear muffs had been provided, but little effort was made to ensure their use. It was held that the workmen would not readily appreciate the dangers of injury to their hearing if they did not use the ear muffs, and the employers were liable for failing to take further steps to impress on the plaintiff the need to use the protective equipment.

9.40 (4) When the employer has done all he can do (and in the context of safety and health, this means doing a great deal), when he has laid down a safe system, provided the necessary safety precautions and equipment, instructed on their use, advised how they should be used properly, pointed out the risks involved if they are not used, and given constant reminders about their use, then he can do no more, and he will be absolved from liability. Admittedly, this does not solve the problem, which is how to ensure that employees are protected from their own folly. Various ways of dealing with the enforcement of safety rules will be discussed in chapter 10.

9.41 If an employee is working on the premises of another, the employer must still take reasonable care for that employee's safety. There may well be some limits to what he can do, but this does not absolve him from doing what he can. In particular, he must ensure that a safe system of working is laid down, give clear instructions as to how to deal with obvious dangers, and tell him to refuse to work if there is an obvious hazard. In *Wilson v Tyneside Window Cleaning Co* the plaintiff was a window cleaner who had, in the course of his employment over a period of ten years, cleaned certain windows at a brewery on a number of occasions. One day he pulled on a handle, which was rotten, and he fell backwards, sustaining injures. His claim against his employers failed. He knew the woodwork was rotten, and he had been instructed not to clean windows if they were not safe. This may be contrasted with *General Cleaning Contractors Ltd v Christmas*, where in almost identical circumstances, the employee succeeded in his claim. The distinction appears to be that in *Christmas*, the employers provided safety belts, but there were certain premises where these could not be used, and there was a failure to instruct the employees to test for defective sashes before the work could proceed. Further, the employers could have provided ladders, or taken other steps to ensure that the work could be performed safely.

iii *Reasonably competent fellow employees*

9.42 If an employer engages an incompetent employee, whose actions injure another employee, the employer will be liable for a failure to take reasonable care. In *Hudson v Ridge Manufacturing Co Ltd* an employee was known for his habit of committing practical jokes. One day he carried one of his pranks too far and injured a fellow employee. The employer was held to be liable. The practical answer in such cases is, after due warning, to firmly dispense with the services of such a person, for he is a menace to himself and to others. On the other hand, an employer will not be liable if he has no reason to suspect that practical jokes are being played, for such acts are outside the scope of the employee's employment, and not done for the purpose of the employer's business (*Smith v Crossley Bros*). In *Coddington v International Harvester Co of Great Britain Ltd* an employee, for a joke, kicked a tin of burning thinners in the direction of another employee. The latter was scorched with flames, and in the agony of the moment, kicked the tin away so that it enveloped the plaintiff, causing him severe burns. The employers were held not liable. There was nothing in the previous conduct of the guilty employee to suggest that he might be a danger to others, and his act was totally unconnected with his employment.

9.43 If an employer appoints an inexperienced person to perform highly dangerous tasks, he may be liable if through lack of experience another employee is injured (*Butler v Fife Coal Co*).

9.44 The employer will be liable for the acts of the employee which are committed in the course of his employment and which injure another employee. If an employee is acting contrary to instructions, he may still be acting within the course of his employment if what he is doing is for the purpose of the employer's business. Thus in *Kay v ITW Ltd* a fork lift truck driver found that his path was obstructed by a lorry. Although he was not authorised to do so, he drove it out of the way, and in the process injured a fellow employee. The employers were held liable, for the act was not so extreme as to take it outside the scope of his employment. However, one must draw a distinction between a prohibition which limits the way the work is done, and those which limit the scope of employment. In *Alford v National Coal Board* there was a statutory provision which stated that shots should not be fired by an unauthorised person. An employee, acting contrary to this provision, injured another employee, but the employer was held not liable.

Proof of negligence

9.45 As a general rule, the burden is upon the plaintiff in an action to affirmatively prove his case. In other words, he must show that the defendant owed to him a duty to take care, that the defendant was in breach of that duty by being negligent, and that as a result of that negligence the plaintiff suffered damage. To assist him in such an action, there are a number of rules of evidence and procedure, designed to enable each side to clarify the issues in dispute and to avoid surprises at the actual court of trial. Thus the court may make an order for inspection of the premises or machinery, it can order one party to disclose documents, records, etc to another, it can order that a party should make further and better particulars of his case, and so on. In *Waugh v British Railways Board* the plaintiff's widow sued for damages, alleging that the defendant's negligence caused the death of her husband. She sought the disclosure of an accident report which was prepared by the defendants partly for the purpose of establishing the cause of the accident and partly to assist their legal advisers to conduct the proceeding before the courts. The defendants resisted the disclosure on the ground of professional privilege, but it was held that the report should be disclosed. A document is only privileged if the dominant purpose for making it in the first place was for the purpose of legal proceedings. If the document had a dual purpose, it would not be covered by professional privilege. The implications of this case in practice can be very wide. Thus if an accident report is made as a result of an employee being away from work for more than three days because of an accident, it will not be privileged. If a safety officer makes a report of an accident, it will also not be privileged. However, once legal proceedings have been commenced, or are imminent, a report prepared for the exclusive use of the company's legal advisers would be privileged.

9.46 Further assistance can be obtained from the inspector under section 28(9) of HSWA which enables him to disclose any factual information about any accident to persons who are a party to any civil proceedings arising from the accident.

9.47 Another rule of evidence which may be of considerable assistance to a plaintiff is *res ipsa loquitor* (let the facts speak for themselves) (see chapter 1). This will apply when the circumstances are such that an accident would not have occurred unless there had been some want of care by the defendant. Thus if a barrel of flour fell out of a building and injured a person walking below, the latter would find it extremely difficult to show that someone was negligent

(*Byrne v Boadle*). In practice, he would not need to do so. He would invoke the rule *res ipsa loquitor*; barrels of flour do not normally fall out of buildings unless someone was negligent, and thus the burden of proof is thrown back to the other party to show that he had, in fact, taken reasonable care. *Res ipsa loquitor* is a rule of evidence, not a rule of law. It creates a rebuttable presumption that there was negligence. If the presumption is rebutted by evidence, then the burden of proof is thrown back to the plaintiff to prove his claim in the usual way.

Defences to a common law action

9.48 Only one action may be brought against an employer in respect of injuries which arise out of one incident. The plaintiff, therefore, must plead his case in such a manner that all possible legal headings are covered. Thus, where appropriate, he should claim in respect of a breach of statutory duty and common law negligence, for each is a separate cause of action. The defendant, for his part, must also be prepared to defend each heading where liability is claimed.

9.49 Since there is no automatic right to compensation, the following defences may be raised.

i *Denial of negligence*

9.50 The employer may deny that he has failed to take reasonable care, or claim that he did everything which a reasonable employer would have done in the circumstances (see *Latimer v AEC Ltd*, above). For example, in *Brown v Rolls Royce Ltd* the plaintiff contracted dermatitis owing to the use of an industrial oil. The employers did not provide a barrier cream on the advice of their chief medical officer, who doubted its efficacy. The employers were held not to be negligent in failing to provide the barrier cream. They were entitled to rely on the skilled judgment of a competent adviser, and no more could be expected. Indeed, the medical officer had instituted his own preventative methods, as a result of which the incidence of dermatitis in the factory had steadily decreased.

9.51 An employer can only take reasonable care within the limits of the knowledge which he has or ought reasonably to have. After all, not every firm (particularly the smaller employer) can have available the resources of specialist expertise. Nonetheless, they must pay attention to current literature which may be available, either from their trade or employers' associations or from other sources. In *Graham v CWS Ltd* the plaintiff worked in a furniture workshop where an electric sanding machine gave off a quantity of fine wood dust. This settled on his skin and caused dermatitis. No general precautions were taken against this, although the manager received all the information which was commonly circulated in the trade. It was held that the employers were not liable. They had fulfilled there duty to take reasonable care by keeping up to date with current knowledge, and were not to blame for not knowing something which only a specialist adviser would have known.

ii *The sole fault of the employee*

9.52 If it can be shown that the injury was the sole fault of the employee, again the employer will not be liable. In *Jones v Lionite Specialities Ltd* a foreman became addicted to a chemical vapour from a tank. One weekend he was found dead, having fallen into the tank. The employers were not liable. In *Brophy v Bradfield* a lorry driver was found dead inside a boiler house, having been overcome by fumes. He had no reason to be there, and the employers had no reason to suspect his presence. Again, they were not liable. And in *Horne v Lec Refrigeration* a tool-setter had been fully instructed on the safety precautions to be followed when operating a machine, but was killed when he failed to operate the safety drill. The employers were held not liable, even though they were in breach of their statutory duty to ensure secure fencing.

9.53 If the claim is based on a breach of statutory duty, the employee cannot, by his own actions, put his employer in breach and then try to blame the employer for that breach. Provided the employer has done all that the statute requires him to do, i e provided the proper equipment, given training, provided adequate supervision, laid down safe systems, and so on, there will come a point when the injured workman will only have himself to blame. In *Ginty v Belmont Building Supplies Ltd* the plaintiff was working on a roof. He knew that it was in a defective state, and that he should not work without boards. The employer provided the boards for use, but the plaintiff failed to use them and fell through the roof. It was held that the employers were not liable for his injuries. They had done all they could do, and the accident was the sole fault of the plaintiff.

9.54 In the nineteenth century the courts were inclined to the view that a worker accepted the risks which were inherent in the occupation and had to rely on his own skill and care, but this view was firmly discounted in the leading case of *Smith v Baker & Sons* and the defence of *volenti non fit injuria* (a person consents to the risks of being injured) is no longer applicable. The fact that the employee knows that there is a risk in the occupation does not mean that he consents to that risk because the employer has been negligent in failing to guard against it. This must apply, *a fortiori*, if the employer is under a statutory duty to guard against the risk. However, if the statutory duty is placed on the employee, and he disregards it, the employer is entitled to raise the defence of *volenti*. In *ICI Ltd v Shatwell* the Quarries (Explosives) Regulations 1959 provided that no testing of an electrical circuit for shotfiring should be done unless all persons in the vicinity had withdrawn to shelter. This duty, which was imposed in order to avoid risks from premature explosions, was laid on the employees. The employers had also prohibited such acts. Two employees were injured when they acted in breach of the Regulations and the employees' instructions, and it was held that the employers could successfully raise the defence of *volenti*.

9.55 The payment of 'danger money' to certain types of employees (e g stunt artistes) may indicate that there is an inherent risk in the occupation which cannot be adequately guarded against, but the real question to be asked is, was the employer negligent? Further, the Unfair Contract Terms Act 1977 states that a person cannot by reference to a contract term or prominently displayed notice exclude or restrict his liability for death or personal injury resulting from negligence.

iii *Causation*

9.56 Although the employer may be negligent, it must still be shown that the injury resulted from that negligence. If the injury would have happened had the employer not been negligent, then the breach of the duty to take care has not caused the damage (see *McWilliams v Sir William Arrol*, para 9.15). Where a breach of statutory duty is alleged, the same principles apply. In *Bonnington Castings Ltd v Wardlaw* the plaintiff was subjected to a silicia dust in premises where there was inadequate ventilation. It was held that the fact that there was a breach of statutory duty (to provide ventilation, Factories Act, section 4), and the fact that the employee suffered from pneumoconiosis did not by itself lead to the conclusion that the breach caused the injury. There must be sufficient evidence to link the one with the other. However, a more liberal view was taken in *Gardiner v Motherwell Machinery and Scrap Co* (see chapter 1) where the court took the view that evidential presumptions may arise in such cases.

iv *Contributory negligence*

9.57 This defence is based on the Law Reform (Contributory Negligence) Act 1945 which provides that if a person is injured, partly because of his own fault, and partly due to the fault of another, damages shall be reduced to the extent the court thinks fit, having regard to the claimant's share in the responsibility for the damage. This defence is successfully raised in many cases. The employer will argue that even if he were negligent, the employee failed to take care for his own safety, and the court may well decide to reduce the damages awarded. Ther is no scientific basis for determining the percentage reduction, and appeal courts may well take a different view of the apportionment of the blame between the parties.

Limitations of actions

9.58 Any action for personal injury or death must be commenced within three years from the date of the accident (Limitation Act 1980, section 11). This means that the actual writ must be issued within the three-year period, although the date of the court hearing may be considerably delayed thereafter. In accident cases, the date of the incident is usually ascertainable, but in some circumstances, where the injury is a result of a constant exposure to the hazard, e g noise which causes deafness (*Berry v Stone Manganese*) or exposure to dangerous substances which can cause cancer (*Wright v Dunlop Rubber Co Ltd*) or pneumoconiosis (*Cartwright v GKN Sankey Ltd*) it is not possible to fix a date, or the plaintiff will be unaware of the date. In such circumstances, the Limitations Act 1980, section 14 provides that the three year period shall begin to run from the date on which the cause of action accrued or the date when the plaintiff had knowledge of the fact that the injury was significant, and that this was due to the employer's negligence. Knowledge, in this connection, means actual or constructive knowledge, i e knowledge which the plaintiff had or ought to have had. For example, if he has received medical advice which indicated an injury, he should know of the likely cause. The Act permits the court to exercise its discretion and allow the action to proceed even though it is outside the limitation period, if it is equitable to do so, having regard to the factors which brought about the delay, the effect it may have on the credibility of witnesses after such a period of time, the disability suffered by the plaintiff, and whether he acted promptly once he realised that he may have a cause of action having regard to any expert advice he may have received.

Damages

9.59 If an employee is killed in the course of his employment, an action may be brought by his personal representatives for the benefit of the estate of the deceased, under the provisions of the Law Reform (Miscellaneous Provisions) Act 1934. Damages will be awarded under the following heads:

(a) loss of expectation of life (this is usually a fairly modest sum);
(b) pain and suffering (if any) up to the time of death;
(c) loss of earnings up to the time of death.

9.60 A further action may be brought simultaneously by his dependants for their own benefit under the Fatal Accidents Act 1976 based on the loss of financial support suffered by, e g his wife, children, or other dependants. Any other money due to the estate is not taken into account, e g personal insurances, pensions payable to the widow etc, but any award made under the 1934 Act will be taken into account. In other words, the one action is brought for the benefit of the deceased estate, the other for the benefit of his family dependants. The two claims are invariably settled together.

9.61 If a person is injured, he will bring the action for his own benefit. Damages will be awarded under the headings of pain and suffering and loss of earnings. If he has received certain social security benefits (see below) one-half will be taken into account and deducted from his tort damages for a maximum period of five years.

9.62 The quantum of damages awarded is frequently a matter of speculation. No amount of money can compensate for the loss of a faculty (arm, leg, eyesight etc) but the courts must try to do their best, and awards will take account of inflation, the permanency of the injury, the effect of earning capacity, additional expenses incurred, and so on.

B State insurance benefits

9.63 An employed person who is injured at work may make a claim for financial assistance under the scheme which is now contained in the Social Security Act 1975 and the Regulations made thereunder. The right to claim is independent of any right of action which may or may not exist at common law for negligence or breach of statutory duty, or indeed any other remedy, for the state scheme is a form of insurance policy, paid for partly by contributions from the employer and employee. Further, questions of fault, blame, contributory conduct, etc are irrelevant to the issue of claiming benefits, provided the employee is within the scope of the relevant provisions.

9.64 The basic outline of the scheme is to provide compensation for a person who suffers a personal injury caused by accident arising out of and in the course of his employment (section 50), or a prescribed disease or personal injury due to the nature of that employment (section 76). Provided certain criteria are met, the claimant will receive a weekly sum in respect of each week's consequent unemployment, up to a maximum of six months. If there is a long term loss of a faculty, he may be able to claim a permanent disability pension. In appropriate cases, death benefit is payable.

INDUSTRIAL INJURIES

9.65 The following criteria must be met:

9.66 (a) There must be a personal injury. This can mean physical or mental injury, such as nervous shock (*Yates v South Kirkby etc Collieries*). However, physical damage to one's property (e g spectacles, clothing, etc) is not covered by the scheme.

9.67 (b) The injury must have been caused by an accident. This word is not defined, and a general commonsense approach must be used. An accident may occur even though the injury was deliberately caused and intended, at least, if one looks at the incident from the point of view of the victim rather than the perpetrator. Thus, in *Trim Joint District School Board v Kelly* a schoolmaster was killed by the deliberate act of some pupils, and this was held to be an 'accident'. An accident is 'an untoward event, which is not expected or designed' (*Fenton v Thorley & Co Ltd*). This must be distinguished from a process which takes place over a period of time (which, though not an accident, may well be a prescribed industrial disease). Thus the gradual contraction of an illness is not an accident, as, for example, when a doctor contracts tuberculosis as a result of a series of penetrations, though a series of events taken together may constitute an accident, e g several minor burns which cumulatively produce a major injury. It will be clear that the borderline between some of the cases is extremely thin and undefined.

9.68 (c) The accident must have arisen out of, and in the course of, the employment. This phrase has also given rise to difficulty. 'In the course of employment' refers to the elements of time (section 52(1)(b)), place of work and job contents. The accident must have occurred after the employment has commenced and before it has terminated, but it is not necessary that the injured person shall actually be working. Thus he is still in the course of his employment if he is having a meal break, or making preparations to start or leave work. If an employee is injured whilst at his normal place of work, he is generally within the course of his employment e g a safety representative performing his functions as such. Some difficulty is experienced with those cases where the employee works in a public place, for the object of the statute is not to compensate for risks common to the general public. Thus if a salesman is injured in a car accident while travelling around to made calls on customers, he would not be entitled to claim. A person is still within the course of his employment if he is doing something which is reasonably incidental to that employment, e g stopping to talk to a fellow worker. However, in *R v National Insurance Commissioner, ex parte Michael*, a policeman was injured while playing football for his local police force. Such activities, though not within his contractual obligations, were encouraged, but it was held that playing football was not part of his ordinary work, and was not reasonably incidental thereto.

9.69 The next requirement is that the accident must arise out of the employment. This represents the casual link between the two. 'Was it part of the injured man's employment to hazard, suffer or to do that which caused his injury?' (*Lancashire and Yorkshire Rly Co v Highley*). The decisions in this branch of the law are equally disparate, as each case will be determined on its own facts. Thus an injured employee is not disbarred from claiming in respect of his own negligence or misconduct, unless his action created a risk which was entirely different from that which was inherent in his employment in normal circumstances. Further, there is a statutory presumption that if an accident arises in the course of the employment, it is deemed to have arisen out of the employment in the absence of evidence to the contrary (section 50(3)).

9.70 In addition, the Social Security Act deals with four special situations.

9.71 (1) An accident shall be deemed to arise out of the course of the employment even though at the time of the accident he was acting contrary to a statutory provision or a Regulation or orders given by his employer, or is acting without instructions from his employer, provided the accident would still have been deemed to arise out of his employment had he not been acting in contravention of the law or his employer's instructions, and the act was done for the purpose of the employer's trade or business (section 52). Thus if a workman is doing something he was forbidden by law to do, or not authorised to do, he may still be entitled to benefit provided, in the absence of such legal rule or instruction, he was in the course of the employment, and acted for the purpose of the employer's business.

9.72 (2) An accident which occurs while the employee is travelling to or from work (with the express or implied permission of the employer) as a passenger in a vehicle, shall be deemed to arise out of and in the course of the employment, notwithstanding that the employee was under no obligation to use the transport, if the accident would have been deemed to have so arisen had he been under such an obligation, and the vehicle is being operated by or on behalf of the employer (or some other person by arrangement with the employer), and it is not being operated in the ordinary course of public transport. In other words, the employee may be entitled to benefits if he is injured when using transport provided by his employer (section 53).

9.73 (3) An accident which happens while the employee is at any premises at which for the time being he is employed for the purpose of his employer's trade or business shall be deemed to arise out of and in the course of the employment if it happens while he is taking steps on an actual or supposed emergency, and he is trying to rescue, succour or protect persons who are in peril, or while he is trying to avert or minimise serious damage to property (section 54).

9.74 (4) An accident shall be treated as arising out of the employment, if it arises in the course of the employment, and it is caused by another person's misconduct, skylarking or negligence, or the behaviour or presence of any animal, or he is struck by lightning, and the employee did not directly or indirectly induce or contribute towards the happening (section 55).

PRESCRIBED DISEASES AND INJURIES

9.75 A person who suffers from a prescribed disease or injury, which is a disease or injury due to the nature of the employment, is also entitled to claim benefits under the Act. The disease or injury must be one which has been prescribed as such by the Secretary of State, if he is satisfied that it ought to be treated as a risk of the occupation (and not just a risk common to all employments) and that it is attributable to the nature of the employment (section 76).

9.76 There is a presumption that if the employee works in the prescribed employment, and he contracts the prescribed disease or suffers the prescribed injury, the disease or injury will be regarded as being due to the nature of the employment unless the contrary is proved (section 77).

9.77 The Secretary or State may also make special provision by Regulations for cases of pneumoconiosis which is accompanied by tuberculosis, emphasema and chronic bronchitis, and for occupational deafness and byssinosis. For a list of prescribed industrial diseases and injuries, see Appendix D.

Benefits under the Act

i *Injury benefit (section 56)*

9.78 This is payable for every day during the period in which the employee is unable to work due to the injury. The payment is a flat rate with various increases for dependants, and the amount payable for each day's absence shall be 1/6th of the weekly rate. The injury benefit period is 156 days (excluding Sundays) beginning with the date of the accident. Absence for a single day is not included, and the first three days' absence are excluded.

ii *Disablement pension and gratuity (section 57)*

9.79 An employee who loses the use of a physical or mental faculty (exceeding an assessment of 1 percent), including a disfigurement (whether or not accompanied by the loss of a faculty) is entitled to claim a disablement pension or gratuity. If the disablement is assessed at less than 20 percent the benefit payable will be in the form of a gratuity, which will be a fixed lump sum payable in one or more instalments. If the disablement is assessed at more than 20 percent, the benefit will be a pension, calculated at a tariff which is set out in the Regulations.

iii *Unemployability supplement (section 58)*

9.80 If the degree of disability is such that the employee is permanently unable to work, an additional unemployability supplement is added to the disability pension.

iv *Special hardship allowance (section 60)*

9.81 If the employee is permanently incapable of following his regular occupation, and is incapable of following employment of a similar standard which is suitable in his case, or because of the loss of a faculty he is incapable of following any occupation or employment, then a special hardship allowance is payable. This is not payable if he is in receipt of the unemployability supplement.

v *Constant attendance allowance (section 61)*

9.82 If a disability pension is assessed at 100 percent, and because of the loss of a faculty the beneficiary requires constant attendance, a constant attendance allowance shall be payable.

vi *Hospital treatment (section 62)*

9.83 Where a person is in receipt of treatment as an in-patient in a hospital or similar institution, his disability shall be treated as being 100 percent for the purpose of his disability pension.

vii *Exceptionally severe disablement (section 63)*

9.84 Where a person is in receipt of a disability pension, and is in receipt of a constant attendance allowance (or would be but for the fact that he is in a hospital or other institution), and the need for constant attention is likely to be permanent, an additional increase in the pension is payable.

viii *Dependent children's allowance (section 64)*

9.85 For a person who is in receipt of injury benefit (i e (i), above) or in receipt of a disability pension together with an unemployability supplement (i e (iii), above), the pension shall be increased in respect of each child of the family.

ix *Adult dependant's allowance (section 66)*

9.86 The weekly rate of injury benefit and disablement pension with an unemployability supplement shall be increased where the beneficiary is living with his wife or contributing towards her maintenance, or, if a woman has a husband who is incapable of supporting himself, or the beneficiary is supporting a prescribed relative who lives with him, or there is a woman who is looking after the beneficiary's children.

Industrial death benefits

i *Widow's benefit (section 67)*

9.87 The widow of a deceased is entitled to receive a weekly pension if at the time of the deceased's death she was living with him, or being maintained by him. The pension is payable for life or until she remarries. However, she is not entitled to the pension if she cohabits with a man who is not her husband.

ii *Widower's benefit (section 69)*

9.88 A widower is entitled to receive death benefit if at the date of his deceased wife's death he was wholly or mainly maintained by her, and was permanently incapable of self-support.

iii *Children's benefits (section 70)*

9.89 A person who looks after the children of the deceased is entitled to an appropriate allowance. If the person is the widow of the deceased, the allowance is somewhat higher than if it is some other person.

iv *Parents' allowance (section 71)*

9.90 A parent of the deceased shall be entitled to death benefit if at the deceased's death the parent was to a substantial extent being maintained by the deceased, or would, but for the accident have been so maintained.

v *Dependent relative allowance (section 72)*

9.91 Any prescribed relative shall be entitled to death benefits if at the deceased death he/she was wholly or mainly maintained by the deceased at his death, or was being so maintained to a substantial extent, and the relative (if a man) was permanently incapable of self-support, or (if a woman) was incapable of self-support or living with a husband who was so incapable.

vi *Woman caring for children of the deceased (section 73)*

9.92 A female who was residing with the deceased and had care of his children shall be entitled to death benefits if she was wholly or mainly maintained by him.

Claims for benefit

9.93 These will be dealt with initially by the local insurance officer. There is a right of appeal from his decision to the local appeal tribunal, to which matters may also be referred by the insurance officer. An appeal will then lie to the Commissioner. Certain matters may be referred by the insurance officer direct to the Secretary of State or to the Attendance Allowance Board for a determination.

9.94 Questions which relate to disablement pensions, such as whether or not the applicant has lost a faculty, or the degree of disability, are referred to the medical board, with a right of appeal to a medical appeal tribunal.

Reporting accidents by employees

9.95 Under the Social Security (Claims and Payments) Regulations 1979 every employee is required to report an accident which results in a personal injury to himself and in respect of which he intends to claim industrial injuries benefits. The report may be made orally, or in writing, but must be made as soon as is practicable after the accident has occurred. The report may be made by the employee, or by someone acting on his behalf, and it may be given to the employer or some person acting on his behalf, or to his foreman or supervisor. Form BI 510 may be used for this purpose.

C Other legal remedies

9.96 An employee who is injured by an act which amounts to a criminal offence on the part of some person may find that he is unable to sue his employer for compensation, for the employer will not have broken any legal duty towards him. However, in 1964 Parliament created the Criminal Injuries Compensation Board, which may make *ex gratia* payments in such circumstances. These arise when a person suffers personal injury directly attributable to a crime of violence, or attempting to arrest an offender. The injury must be one for which compensation of more than £250 will be awarded, the circumstances of the incident must be reported to the police, and the applicant has given all reasonable assistance to the Board, particularly in relation to any medical reports which may be required. Injuries caused by traffic offences are excluded from the scheme, unless there is a deliberate attempt to run the applicant down.

9.97 By way of example, we may cite the case of *Charlton v Forrest Printing Ink Co*, where the plaintiff was a manager at the defendants' works. One of his duties was to collect the wages from the bank. In 1974 there was an unsuccessful attempt to snatch the wages, and thus the managing director gave instructions to those who were to collect the wages that they should take precautions, such as varying the route taken each week, using different modes of transport, and sending different people. Despite these instructions, a pattern of collection tended to develop. In 1977, when the plaintiff was returning from the bank, he was attacked by bandits who threw ammonia into his face, causing severe damage to his eye-sight. He claimed damages from his employers, arguing that they had failed to take reasonable care for his safety. In the High Court, the judge accepted the argument that although the sum of money involved was only small (£1,500), in view of the previous robbery, it would have been reasonable to use a professional security firm for this task, and hence the employers were in breach of their common law duty to take sensible precautions to protect employees who were involved in a hazardous task. This decision was reversed by the Court of Appeal. Proper steps had been taken to instruct the employees to vary their methods used for collecting the wages. It was unreasonable to expect the employers to guard against a possibility which would not influence the mind of a reasonable man. Statistics showed that the majority of firms of this size did not employ specialist security firms. Hence, the employers were not negligent. However, Lord Denning went on to point out that the applicant would not doubt have a good claim if he applied to the Criminal Injuries Compensation Board.

9.98 Another type of scheme is operated by the Motor Insurers' Bureau, which was set up in 1969 by an agreement between insurance companies and the Ministry of Transport. If a person suffers death or injury arising out of the use of a motor vehicle, and he is unable to trace the person responsible, or the person responsible is unable to pay compensation (usually because he is uninsured), then the Board will accept the liability to compensate the injured party, the damages to be assessed in a like manner as a court applying the normal legal principles would assess. In practice, the Bureau nominates an insurance company to accept the risk, the latter will then seek to recover the damages paid (in so far as they are able to do so) from the wrongdoer.

9.99 The Pneumoconiosis etc (Workers' Compensation) Act 1979 provides for lump sum payments to be made to persons who are disabled by industrial lung diseases (pneumoconiosis, byssinosis and diffuse mesothelioma). The conditions are that the claimant must be entitled to disablement benefit under the Social Security Act 1975 (para 9.63, above), he must be unable to recover damages from his employer because the latter has gone out of business and no legal action has been brought or compromised. This scheme is an interesting example of the state providing compensation to injured workers in circumstances where tort liability claims would not be met. A similar scheme for victims of pneumoconiosis has been in force on a voluntary basis for coal miners since 1974, and is operated by the National Coal Board.

9.100 Finally, under the Powers of Criminal Courts Act 1973 the criminal courts (including magistrates' courts) are empowered to make an award of compensation to a person who has been injured or suffered any loss or damage as a result of the commission of a criminal offence. This is in addition to any other punishment which the court may impose. The maximum amount of compensation which may be awarded is £1,000. Thus, in strict theory, if an employer was prosecuted for an offence under HSWA or other legislation for an offence which caused personal injury or damage to an employee's property, the latter could apply to the court for a compensation order to be made at the same time. Indeed, the court can make such an order on its own volition. If the injured person subsequently brings civil proceedings, any compensation awarded in the criminal court will be taken into account. Although this is a useful provision for dealing with claims in respect of minor personal injuries, very little use appears to have been made of it.

10 Enforcing health and safety rules

10.1 The duty to ensure health and safety policies are observed falls on all those who are involved in the work processess – on the employer, management, safety officer, shop steward, safety representative, supervisor, and employees generally. Each must play his own special role in ensuring that health and safety policies are laid down, adequately promulgated, and, perhaps the most important of all, carried out in practice.

10.2 The basis for these duties is the contract of employment, which is essentially mutual agreement. It follows that a breach of contract by either side entitles the other party to pursue whatever remedy is appropriate in the circumstances. Thus, on the one hand, if the employee commits a breach of contract, the employer may terminate the contract, and, in theory – at any rate – sue for any damage he has suffered. On the other hand, there is an implied term of the contract that the employer will take reasonable care to ensure the health and safety of his employees while at work (*Matthews v Kuwait Bechtel Corporation*), and if he is in breach of that term the employee may 'accept' the breach and resign (this is the doctrine of constructive dismissal). If he has suffered damage, he may claim in respect of these. Additionally, if the employer's breach has caused injury, he may sue in tort for a breach of duty imposed by law (see chapter 9).

Appointment procedures

10.3 Health and safety policies begin with appointment procedures. In theory, the employer should ensure that all his employees are fit, healthy, with 100 percent vision and hearing capacity, and not suffering from any disability which renders them liable to an accident or to make them a hazard to themselves or to others. To this aim, pre-employment medicals may be used. The policy is as foolish in practice as it is sound in theory. By the time pre-medical screening has eliminated all those who are not suffering or who have never suffered from some medical disability, either there will be no-one left to employ or employment will be limited to a few Amazons!

10.4 Pre-medical screening must be related to a particular hazard. Judgment must be made on medical and safety grounds, not on the ground of general health. Epileptics can do a useful job of work, depending on the degree of their illness and the nature of the danger they face at work. Disabled employees can do work other than act as car park attendants, if the work is matched to their abilities, and care is taken to ensure their safety. Dedication to health and safety does not mean the exclusion of a large section of the population from employment opportunities. Thus the object of pre-medical employment must not be to exclude certain people, but to place employees in the appropriate niche.

10.5 Nonetheless, critical situations may require extreme measures. In *Jeffries v BP Tankers Ltd* it was the company's policy not to employ people as radio operators on ships if they had had any history of cardiac disease. They dismissed a radio officer who had suffered a heart attack, even though he had made an excellent recovery. His dismissal was held to be fair, for the policy was one which had to be rigorously enforced for reasons which the employers were satisfied were necessary.

10.6 Health considerations are equally relevant. In *Panesar v Nestle Co Ltd* the applicant, who was a Sikh, enquired about a job with the respondents. He was told that there was a company rule prohibiting beards, and since he was not prepared to shave off his beard, he was not interviewed for the job. He claimed that he had been unlawfully discriminated against on grounds of race (see Race Relations Act 1976, section 4). It was conceded that he had been indirectly discriminated against, in that the proportion of applicants of his race (assuming that Sikhs are a racial group) who could comply with the requirement that they should not wear a beard was smaller than the proportion of applicants from other racial groups (Race Relations Act, section 1(1)(b)). However, the requirement or condition was capable of being justified on grounds other than race, namely hygiene considerations, and his claim failed before the industrial tribunal, the EAT and the Court of Appeal. A similar finding would doubtless apply if it was a condition of employment that an employee should wear a safety helmet, and provided it could be shown that there was a hazard which needed to be guarded against, and that the policy was rigorously enforced without exception, this would not amount to unlawful discrimination against Sikh applicants.

Works rules

10.7 A rule book is a valuable source of information and guidance, and should in particular lay down precise rules concerning all aspects of health, safety and welfare. The contents of the rule book should be incorporated into the contract of employment, they should be up-dated in accordance with developments and experience, and should be contained in a booklet which can be retained by the employee. Somewhat strangely, section 1 of the Employment Protection (Consolidation) Act (which requires employers to give to their employees written particulars of their terms and conditions of employment) includes a requirement to provide information relating to disciplinary and grievance procedures, but specifically states that this does not apply to rules, disciplinary decisions, grievances or procedures relating to health or safety at work. This exclusion is even more surprising when it is realised that health and safety matters probably constitute the biggest single cause of disciplinary problems. It is to be hoped that this gap in the law is remedied in the near future. However, the Code of Practice issued by the Advisory, Conciliation and Arbitration Service (ACAS) entitled 'Disciplinary Practice and Procedures in Employment' states that 'When drawing up rules the aim should be to specify clearly and concisely those necessary for the efficient and safe performance of work'.

10.8 Particular attention should be paid to publicising the contents of the rule book to new employees, on some form of induction course. Those rules which are mandatory can be highlighted, and disciplinary procedures and sanctions which will follow a breach can be spelt out. Attention can be paid to ways of communicating the rules to immigrant employees and others whose command of English is less than perfect, so that they are unable to plead ignorance. Records can be kept of when and how the rules were communicated to each individual concerned.

10.9 Rules which relate to health and safety are designed to protect employees individually and collectively, and should therefore contain sanctions in the event of a breach. The matter was expressed forcefully by Stephenson LJ in *Bux v Slough Metals Ltd*, when he said 'The employer must make the law of the land the rule of the factory'. In other words, the employee should be informed as to the legal requirements, and warned that he is breaking the law if he fails to wear to use the precautions provided, for which he can be prosecuted. He can also be warned that he faces disciplinary action or even dismissal if he acts in breach of the safety rules or the legal requirements.

10.10 To achieve consistency in procedure, certain principles must be observed. First, the rules must be promulgated. This means that they must be brought to the attention of the employees in a suitable form. It is no longer regarded as being sufficient to post rules on a notice board or hide them in the personnel office. In *Pitts v Rivertex Ltd*, the employee was dismissed for breaking a works rule, a copy of which could be found on the notice board. It was held that if a rule was so important that a breach would be visited by instant dismissal, it should have been expressly communicated to all the employees concerned.

10.11 Second, rules must be reasonable. An employer may impose has own standards, but they must have a sound functional basis, and not be old-fashioned, out-of-date, or based on prejudice. In *Talbot v Hugh Fulton Ltd* the applicant was dismissed for having long hair in breach of the company's rules. The dismissal was held to be unfair. Management did not specify the length of hair which was acceptable, it was not shown that long hair was a safety hazard, and the rule did not appear to apply to women with long hair. Clearly, it was an act of prejudice against the hair style of the day. By contrast, we can cite *Marsh v Judge International*, where a youth had hair which was 2 feet 6 inches long, reaching down to his waist. The factory inspector told the employers that they would be prosecuted if the youth caught his hair in any machinery and was injured, and so, after due warnings to cut his hair (which he ignored) the youth was dismissed. This was held to be fair. The rule, in the nature of an instruction, was perfectly reasonable, and in the interests of safety. It follows that an employer can lay down suitable safety standards provided they are functional and reasonable (*Singh v Lyons Maid Ltd*).

10.12 Third, the rules must be consistently enforced. If a rule is generally disregarded, or no severe sanction is imposed for a breach, then it may be unfair to act on it without giving some indication that there was to be a change in enforcement policy. Thus in *Bendall v Paine and Betteridge* the applicant had been employed for fifteen years. From time to time he was told to put out a cigarette he was smoking, as there was a fire risk. One day, he was summarily dismissed for smoking, and this was held to be unfair. In the past he had been warned without it being brought home to him that he was risking instant dismissal. He should have been given a final, written warning before he was dismissed.

10.13 Health and Safety rules should therefore be drawn up in accordance with the hazards perceived. This requires a careful assessment of the likely risks, based on the nature of the firm, its processes, the workforce, and all other relevant considerations. The following are some examples of the more common rules which should be considered.

A SMOKING

10.14 If there is a risk of fire (*Bendall v Paine and Betteridge*) or a health hazard (*Unkles v Milanda Bread Co*), clear rules should be laid down informing employees that anyone caught smoking in the prohibited area will be instantly dismissed. Areas where smoking is permitted should be clearly defined. Additionally, the employer must pay attention to the statutory provisions which prohibit smoking. These include:

(a)	Celluloid Regulations 1921	no person shall be allowed to smoke in any room in which celluloid is manufactured, manipulated or stored
(b)	Chemical Works Regulations 1922	a prominent notice prohibiting smoking shall be affixed at the entrance of every room in which there is liability to explosion from inflammable gas, vapour or dust
(c)	Control of Lead at Work Regulations 1980	an employee shall not smoke in any place which he has reason to believe is contaminated by lead
(d)	Manufacture of Cinematograph Film Regulations 1928	no person shall take any smoking materials into any room in which cinematograph film is manufactured, repaired, manipulated or used
(e)	Cinematograph Film Stripping Regulations 1939	no smoking materials shall be allowed in any part of the premises

(f) Vitreous Enamelling Regulations 1908	no employed person shall introduce tobacco into any room in which an enamelling process is carried on
(g) Highly Flammable Liquids and Liquified Petroleum Gases Regulations 1972	no smoking shall be permitted when flammable liquids are present. The occupier shall take all reasonable practicable steps to ensure compliance, including the display of a clear and bold notice
(h) Factories (Testing of Aircraft Engines and Accessories) Special Regulations 1952	no smoking shall be allowed in any place where aircraft engines are being tested
(i) Horsehair Regulations 1907	no person shall introduce tobacco in any room in which non-disinfection material is stored or manipulated
(j) Ionising Radiations (Unsealed Radioactive Substances) Regulations 1968	no person shall make use of tobacco while working in an active or contaminated area
(k) Magnesium (Grinding of Castings and other articles) Special Regulations 1946	no smoking shall be permitted within 20 feet of certain processes or in any room where magnesium dust is being kept, or smoke when handling magnesium dust
(l) Pottery (Health and Welfare) Special Regulations 1950	no person who has been employed in certain processes shall make use of tobacco unless he has washed his hands since being engaged on that process
(m) Health and Safety (Agriculture) (Poisonous Substances) Regulations 1975	no worker shall smoke unless he has removed all protective clothing, has washed his hands and face, and is outside the area in which he might be poisoned by any specified substance which has been used
(n) Petroleum Spirit (Conveyance by Road) Regulations 1957	
(o) Organic Peroxides (Conveyance by Road) Regulations 1973	
(p) Explosives Act 1875, section 10	

B EATING AND DRINKING

10.15 The following provisions prohibiting the partaking of food or drink are in force:

(a) Horsehair Regulations 1907	no person shall introduce, keep, prepare or partake of food or drink in any room in which materials are stored or manipulated which have not undergone disinfection
(b) Ionising Radiations (Unsealed Radioactive Substances) Regulations 1968	no person working in an active area etc shall partake of food or drink there (except drink from a fountain which is not contaminated)
(c) Non-Ferrous Metals (Melting and Foundring) Regulations 1962	no person shall be permitted to take a main meal in any indoor workroom where processes are being carried on
(d) Pottery (Health and Welfare) Special Regulations 1950	no person shall introduce food or drink into any place in which a scheduled process is being carried on
(e) Tinning of Metal Hollow-Ware, Iron Drums and Harness Furniture Regulations 1909	no person employed in tinning mounting, denting or scouring shall keep, prepare or partake of any food or alchoholic drink in any room in which such work is being carried on
(f) Control of Lead at Work Regulations 1980	an employee shall not eat or drink in any place which he has reason to believe to be contaminated by lead
(g) Health and Safety (Agriculture) (Poisonous Substances) Regulations 1975	no person shall eat or drink unless he has taken certain precautions
(h) Factories Act 1961, section 64	no person shall be permitted to partake of any food or drink in any room where arsenic or other poisonous substances are used so as to give rise to any dust or fumes.

C FIGHTING

10.16 This can lead to a serious accident occuring, and the rule should lay down that anyone caught fighting will be subject to disciplinary proceedings. It cannot be right to state that fighting will lead to instant dismissal; after all, if two people are fighting, one may be merely defending himself, or the one who started the fight may have been provoked. A careful investigation of the circumstances is therefore called for. Further, two people may be fighting in a no-risk area, which may invoke a disciplinary sanction less than dismissal. The rule, therefore, should indicate the discretionary power of management. In *Taylor v Parsons Peebles Ltd*, the applicant had been employed for twenty years by the company before he was dismissed after he was involved in a fight. It was the company's policy to dismiss automatically any employee who deliberately struck another employee. The industrial tribunal held that the dismissal was fair, because the policy was applied consistently and the other employee who was fighting was also dismissed. On appeal, the decision was reversed by the EAT. In determining whether a decision to dismiss was reasonable, the proper test was not the employer's policy, but what the reaction would be of a reasonable employer in the circumstances. The employers' rules of conduct must be considered in the light of how it would be applied by a reasonable employer. Taking into account the fact that the employee had been employed for twenty years with no serious disciplinary record against him, a reasonable employer would not have applied the rigid sanction of automatic dismissal. In other words, the employer must be consistent in his procedures, but flexible in his punishment. He must consider the gravity of the offence, and the circumstances of the offender. The employee must always be given at least an opportunity to plead mitigating circumstances, although the weight to be attached to such a plea is for the employer to decide, bearing in mind the gravity of the incident. In *Taylor's* case, because the employers had applied a policy consistently without taking into account relevant factors, the dismissal was unfair. However, it was also held that the employee had contributed towards his dismissal, and compensation was reduced by 25 percent.

D DRUNKENNESS

10.17 If an employee is found to be drunk on the premises, or to have been drinking alchohol, it may be good policy to escort him off the premises, for the fact of drinking *per se* may not be sufficient to warrant dismissal (*McGibbon v Gillespie Building Co*). However, disciplinary proceedings should normally be instituted. If the drunkenness is such as to cause or constitute a serious safety hazard, different considerations would apply. In these circumstances, the matter should be dealt with as a serious disciplinary offence, and the procedure should be activated immediately. In

Abercrombie v Alexander Thomson & Son the applicant was found to be in an intoxicated state while in charge of a crane. A decision was taken to dismiss him, but this was not implemented until two weeks had elapsed, during which time he was permitted to carry on working. The dismissal was held to be unfair. By delaying the taking of action, the employers had condoned the offence.

10.18 A person who has symptoms of alchoholism should not be put through the disciplinary procedure immediately. Rather, his condition should be treated as an illness, and medical advice should be sought as to his condition and the likelihood of recovery or obtaining treatment. HSE has produced an occasional paper entitled 'The Problem Drinker at Work' which discusses the benefits which can be obtained from a policy aimed at encouraging problem drinkers to seek assistance and treatment.

10.19 There are some statutory provisions which also need to be borne in mind. These include:

(a) Work in Compressed Air Special Regulations 1958 (no person employed shall consume alcohol whilst in compressed air);
(b) Tinning of Metal Hollow-Ware etc Regulations 1909 (no person employed in tinning etc shall keep or prepare any alcoholic drink in any room in which such work is carried on).

There are a number of Regulations which prohibit the consumption of food and drink whilst engaged in certain processes, and these could certainly cover the partaking of alcohol.

E SKYLARKING

10.20 The dangers from employees who indulge in horseplay or skylarking have already been noted (see chapter 9), and the rules should state that such conduct will not be tolerated. Employees, particularly the young or inexperienced, should be given final written warnings, indicating that any repetition will result in instant dismissal. In *Hudson v Ridge Manufacturing Co* the plaintiff was injured as a result of a prank played upon him by an employee who was known for horseplay. The employers were held to be liable. It was stated that 'If a fellow workman. . . by his habitual conduct is likely to prove a source of danger to his fellow employees, a duty lies fairly and squarely on the employer to remove the source of danger.'

F BREACH OF SAFETY RULES

10.21 Again, this will be dealt with in accordance with the hazard incurred. In *Martin v Yorkshire Imperial Metals* the applicant tied down with a piece of wire a lever on an automatic lathe. This had the effect of by-passing the safety device, for the machine could only be used if the operator used both hands. It was held that his dismissal was fair. He knew that he would be dismissed if he neglected to use the safety device. In *Ashworth v John Needham & Sons* the applicant acted in flagrant breach of safety rules by putting a fence around a hole in the ground instead of replacing a plate. After the matter had been discussed with the company's safety officer and a trade union official he was dismissed. This was held to be fair, as a serious accident could have happened.

G FAILURE TO USE SAFETY EQUIPMENT

10.22 Again, the gravity of the consequences should be a factor which determines the severity of the sanction, and the rules should be flexible enough to deal with this situation. Thus in *Frizzell v Flanders* the employee was provided with a gas mask while working in a tank. He was seen working without the mask and was dismissed. This was held to be fair. It was essential to enforce rigorously the use of safety equipment. On the other hand, in *Henry v Vauxhall Motors Ltd* the employee discovered that his safety helmet was missing. He was provided with another one which, he claimed, was uncomfortable. He therefore worked without the helmet, and persisted in his refusal after being instructed to wear it by the foreman. He was dismissed, but this was held to be unfair. There was no proper enquiry into the reasons for his refusal. Nonetheless, the industrial tribunal thought that he had contributed substantially to his dismissal, and reduced his compensation award by 60 percent.

H HYGIENE OBSERVANCE

10.23 It may be necessary to enforce standards of hygiene because there is a risk of contamination to the product, or a health hazard to employees. For example, in *Gill v Walls Meat Co*, the applicant, who was a Sikh, was employed to work with open meat. He did not have a beard at the time he commenced employment, but at some later stage 'he was converted to the paths of righteousness', donned a turban, and grew a beard. He was told that he could not be employed on that particular job whilst wearing his beard, but he refused to shave it off. He was offered alternative employment where he would be permitted to have his beard, but he refused, and he was dismissed. This was held to be fair. Similarly, the rules should provide that employees who have been suffering from or in contact with a contagious disease should be encouraged to report to the appropriate medical adviser before commencing work, to ensure that they are 'clear' and do not transmit the disease or illness to other employees or do not cause contamination of the product. Personal hygiene is also important. In *Singh v John Laing & Sons* a company rule provided that anyone misusing the toilet facilities would be liable to instant dismissal. The applicant was seen urinating in a room he had been instructed to clean, and he was thus dismissed. This was held to be fair. He knew of the rule, had a good command of English and could offer little by way of explanation. After all, employers are legally bound to maintain toilet facilities in clean and proper condition and thus it is not wrong to enforce high standards of hygiene (*Singh v Elliotts Bricks*).

I HEALTH GENERALLY

10.24 In some circumstances, general ill health must be reported and noted, so as not to cause a hazard. In *Singh-Deu v Chloride Metals Ltd* the employee worked in a lead smelting factory where it was essential to remain alert in view of the dangerous processes which were being carried on. He was sent home by the works doctor after complaining of feeling unwell. He visited his own doctor, who diagnosed paranoid schizophrenia. He was then examined by a specialist, after which he attempted to return to work. However, the works doctor would not allow him to return unless the specialist gave an assurance that there would be no recurrence of the illness. This assurance was not forthcoming, and so he was dismissed, as the works manager, mindful of the inherently dangerous processes being carried on in the lead factory, was not prepared to take the responsibility of allowing him back at work. The dismissal was held to be fair. It was not reasonable to continue to employ him in a delicately balanced job which called for a high degree of concentration. The effect of a relapse on the applicant and on others could have been devastating had a mistake or error been made. A similar result was reached in *Balogun v Lucas Batteries Ltd* where the applicant who was working with lead, suffered from hypertension. Medical evidence clearly indicated that this type of work can be harmful, especially to those who have an existing predisposition to certain types of illnesses, including hypertension. Efforts were made to find him alternative work, but there was no suitable job, and he was dismissed. This was held to be fair. His continued employment constituted a risk to himself and to others.

J VANDALISM

10.25 An employee who misuses the company's property or interferes with anything provided for the use of employees generally can be disciplined. This is especially true if there is an interference or misuse of anything provided for health, safety and welfare, for there is also a breach of section 8 of HSWA. In *Ferodo Ltd v Barnes* an employee was dismissed after the company decided he had been committing an act of vandalism in the lavatories. The industrial tribunal decided that the dismissal was unfair, as they were not satisfied that in fact the employee had committed this act. This was reversed on appeal. It is not the duty of the industrial tribunal to re-try the case. It was their duty to see that management acted reasonably, and as long as there had been a fair and proper investigation, the industrial tribunal should not substitute their judgment for that of management.

K NEGLECT

10.26 An employee who is incompetent or neglectful in his work may cause some damage to the work processes, and the matter should be regarded as a disciplinary problem. The first duty of management is to ensure that the employee has been trained properly, supervised adequately, has sufficient support staff and sufficient facilities to do the job in question. An employee who is suffering from irredeemable incompetence can be offered alternative employment, failing which a dismissal will be fair. If he is suffering from neglectful incompetence, he should be warned, in accordance with the gravity of the case. However, if the neglect is of such a dangerous nature that there is the likelihood of serious injury to persons or damage to property, then the matter becomes a health and safety issue, and can be dealt with as such. This could result in a disciplinary sanction which is appropriate to the case, and even dismissal. In *Taylor v Alidair Ltd*, an airline pilot landed his aeroplane in a manner which caused some concern and consternation among the rest of the crew and passengers. After a full investigation it was decided that he had been negligent, and he was dismissed. This was held to be fair. There are some activities the consequences of which are so serious and grave that it is not possible to risk a repetition. Nor does it matter if the neglect causes a risk of injury to fellow employees or to the public or customers of the employer. Thus if an employee fails to follow the prescribed safety checks (*Wilcox v HGS*) or the work is done in a negligent manner which creates a risk (*McGibbon v Gillespie*) he may be dismissed.

L SLEEPING ON DUTY

10.27 Whatever may be the position in other work situations, it has always been recognised that safety considerations must be regarded as being paramount. In *Jenkins v British Gypsum Ltd* the applicant was employed on the night shift checking and taking the temperature of a gas fired kiln. He was found asleep on duty, and was dismissed, even though there was no specific rule to cover this situation. Nonetheless, his dismissal was held to be fair. It was essential for him to monitor regularly the temperature as a safeguard, and the fact that he had to be aroused in order to do his work constituted gross misconduct. 'Alertness is essential from the safety angle' commented the industrial tribunal.

M TRAINING

10.28 Since the employer is under an obligation to train his employees in safety and health matters, and since also the employee is obliged to co-operate with the employer in the performance of the statutory duties, an employee who refuses to be trained may be fairly dismissed. In *Minter v Willingborough Foundries Ltd* the applicant was a nurse employed at the employer's factory. There were complaints about her standards of medical care, and she was asked to go on a training course, but she refused. Subsequently, there were further complaints, and she was again asked to undertake further training, and again she refused. Because the employers were concerned about their obligations under HSWA, she was dismissed. The EAT upheld the finding of the industrial tribunal that the dismissal was fair. The course would not have involved her in any expense or inconvenience, and she did not give an adequate explanation for her refusal.

N OTHER DANGEROUS PRACTICES

10.29 No list of actions which are to be the subject of disciplinary sanctions can be exhaustive, and each employer must try to complete the list in accordance with his own situation, as well as covering the unexpected. Thus, it should be fairly obvious that it is an extremely dangerous practice to light a bonfire near to flammable material (*Bussey v CSW Engineering Ltd*) or to fire air guns whilst at work, even though this is during the lunch hour (*Shipside (Ruthin) Ltd v T&GWU*). To smoke in a wood and paint shop (*Bendall v Paine*), to walk out, leaving a high pressure steam boiler on (*Gannon v Firth*), to drive a vehicle without being qualified to do so, or to drive a vehicle so badly that the brakes overheat

(*Potter v Rich*), are all examples of conduct capable of attracting disciplinary sanctions, including dismissal. The list can be extended almost indefinitely.

Enforcing safety rules: action by the employee

10.30 If there is an actual injury to an employee, he can, of course, pursue whatever remedy is available to him at common law (see chapter 9). If there is a threatened injury, the situation is more delicate. Since it is an implied term of the contract of employment that the employer will ensure the health and safety of his employees, the employer will be in 'breach' of contract if he fails to take the necessary steps. In these circumstances, the employee will 'accept' the breach, and resign, but, in law, since he was entitled to resign by virtue of the employer's conduct, it is the employer who has 'dismissed' the employee. In technical terms, even though the employee has resigned, he may bring a claim for 'constructive dismissal', and, provided he has the requisite period of continuous employment (two years in respect of employers who employ less than twenty employees, one year in respect of all other employees), he can bring his claim before the industrial tribunal for compensation. For example, in *British Aircraft Corporation v Austin* the employee asked her supervisor for a pair of prescription safety glasses. After waiting several months, during which time she heard nothing further, she resigned, and claimed constructive dismissal. Her claim succeeded. It was an implied term of the contract that the employers would ensure her safety, and by failing to investigate her request and provide the necessary safety precaution, they had broken the contract. It will be noted at this stage that the inaction was the fault of the supervisor, yet it was the employer who was held responsible.

10.31 A constructive dismissal claim does not need to be based on an actual injury (in theory, there is no reason why this should not be so, but the author knows of no instance where this has been done); the fear of the possibility is sufficient. In *Keys v Shoefaye Ltd* the employee worked in a shop which was robbed in daytime by a gang of youths. The manager was asked by the employee to do something about the security of the premises, but he replied that there was nothing he could do. A further daytime robbery took place, and the employee resigned, and claimed constructive dismissal on the ground that the premises were no longer safe to work in. Her claim succeeded. The employers were obliged to take reasonable steps to operate a safe system of work and to provide safe premises. The industrial tribunal thought that it might have been possible to instal a telephone, or to employ a male assistant, etc. (It may be doubted whether the first suggestion would be effective, or whether the second would be lawful, but this merely illustrates the unworldliness of some industrial tribunals.)

10.32 However, the contractual obligation of the employer is no higher than the duty owed in tort, which is to take reasonable care only. The employer does not guarantee absolutely the safety of his employees. In *Buttars v Holo-Krome Ltd* the employee was injured when a blank flew out of a machine and struck his safety glasses. A lens broke and injured his eye. When he returned to work he asked for a guard to be fitted to the machine, but the employers claimed it was safe. He resigned and claimed constructive dismissal, but the claim was rejected. The accident had been reported to the factory inspector who agreed that the machine was safe, and proposed to take no further action on the incident. There was no duty on the part of the employer to fence, for this duty does not apply to parts of the machine or materials used by the machine which fly out (*Nicholls v Austin*, see chapter 4). The accident was a freak one, it was not usual to fit guards on this type of machine. In these circumstances, the industrial tribunal thought that the employers had not broken the contract, and hence there was no constructive dismissal.

10.33 More difficult is the situation where the employee responds to the employer's breach by action other than resignation. In *Mariner v Domestic and Industrial Polythene Ltd*, some workers discovered that the temperature in the workplace was 53 degrees. There was no fuel left for heating, as the employers had allowed his supplies to run down in anticipation of the warmer weather. The workers therefore went home, and the following day, when they reported for work, they were dismissed for going on strike. In law, a complaint of unfair dismissal, brought about because a person is taking part in a strike or other industrial action cannot normally be entertained by the industrial tribunal (EPCA, section 62), but in this case, the industrial tribunal decided that the workers had not been on strike, for they had not withdrawn their labour in breach of their employment contracts. It was the employer who was in breach, for he had allowed the temperature to fall below the statutory minimum, and there is an implied term in the contract of employment that the employer will perform his statutory obligations. If he fails to do so, the employees are merely responding to the breach, not acting themselves in breach.

10.34 It is sound practice for an employee to take matters up with the appropriate level of management before taking the drastic step of resignation, but there appears to be no legal reason why he should do so, for the fact that the employer is in breach of contract should be sufficient (*Seligman and Latz Ltd v McHugh*). Thus in *Graham Oxley Tool Steels Ltd v Firth*, the employee had to work in a small bay near to an open door. The only heating came from a radiant heater fixed to the ceiling. One day, when the weather was very cold, she was kept waiting outside the entrance to the premises, and eventually she decided to go home. She subsequently resigned and claimed constructive dismissal. She had made no previous complaint about the cold working conditions, although a subsequent visit from the factory inspector revealed that the temperature was 49 degrees. Her claim for unfair dismissal succeeded. The employer was in breach of his obligation to provide a proper working environment, and the failure to do so constituted a fundamental breach of contract which entitled her to resign.

10.35 Nonetheless, a well-drawn-up grievance procedure which is incorporated into the contract may prove to be a useful method of preventing claims of this nature, particularly in respect of those 'innocent' breaches, about which the employer knows nothing, or which were accidental or unintentional. Thus an employee who fails or refuses to use that procedure may find that he is not entitled to claim constructive dismissal, or, if he does succeed, he may find that his compensation award is reduced, on the ground that by failing to adopt the grievance procedure, he contributed towards his own dismissal, or failed to mitigate against his loss.

10.36 It should finally be pointed out that some of the decisions of the industrial tribunals on constructive dismissal are among the weirdest and most unreliable in employment law generally, and care should be taken not to elevate them into principles of law. The original test formulated in *Western Excavating (EEC) Ltd v Sharp* by the Court of Appeal was, has the employer broken the contract? The more modern approach appears to be, did the employer evince an intention to break the contract? This is slightly different, and neurotic employees who imagine that every little thing which goes wrong in their daily employment automatically gives them a right to claim constructive dismissal should be cautioned about such a false assumption. Further, it must be borne in mind that while constructive dismissal is, in law, a 'dismissal', it is not necessarily an unfair dismissal (*Industrial Rubber Products v Gillon*).

Enforcing safety rules: action by the employer

10.37 Safety rules can be enforced within the context of existing disciplinary procedures. These may be drawn up by management, in consultation with the trade unions or workforce if possible, without their co-operation or agreement if necessary. Details should be given to each employee explaining the steps to be followed, the sanctions which may be applied in accordance with the gravity of the case, and the method of appeal. Further reference should be made to the ACAS Code of Practice on Disciplinary Procedure and Practice.

10.38 A disciplinary procedure should have five characteristics:

10.39 (a) There must be a full and proper investigation to the incident. This should be undertaken as soon as possible (*Abercrombie v Thomson & Son*), and consideration should be given to a short period of suspension (with or without pay, in accordance with the contract and/or procedure) pending such investigation.

10.40 (b) The offender must be told of the charge against him. It is no bad thing to put this in writing, particularly if his command of English is weak, or the charge is a serious one, so that he can obtain advice from any available source.

10.41 (c) He should be given an opportunity to state his case, to plead that he didn't do it, or if he did, it was not his fault, or if it was, there were mitigating circumstances which ought to be taken into consideration, etc.

10.42 (d) He should be given the opportunity to be represented, if he so wishes, by a trade union official or shop steward, by a fellow employee, or by anyone who is willing to speak on his behalf.

10.43 (e) He should be given the right of appeal, to a level of management not previously involved in the decision-making process.

10.44 Obviously, the nature of the disciplinary procedure will vary with the size and resources of the firm. One does not expect the same formalities in a small firm as might exist in a large firm. Equally, the sanction which will be imposed will depend on the nature and seriousness of the offence, the circumstances of the individual, and so forth. The purpose of disciplinary sanctions is to improve the conduct of the offender, to deter others from doing the same or similarly wrongful acts, and to protect the individual, other employees, the public and ultimately, the employer. Thus no simple pattern emerges; sometimes the sanction will be corrective in nature, some times it will be designed to encourage others not to break the rules. The gravity of the sanction will reflect the objectives to be achieved. In the case of a minor offence, a minor sanction will be imposed, such as a warning, which may be verbal. A repetition of the offence, or a different kind of offence, or a serious offence, would be dealt with by a written warning. which should detail the offence, warn as to the consequences which may flow from a repetition of the offence, or any other offence of a similar or dissimilar nature. A very serious offence should be dealt with by a final warning, or by dismissal.

10.45 Depending on the nature of the offence and the circumstances of the offender, other sanctions might be imposed. Thus, consideration could be given to a period of suspension without pay (provided the disciplinary procedure confers this power), a transfer to other work, or even a 'fine' (e g if an employee is failing to use safety equipment), which has been agreed as a recognised type of punishment (perhaps with the proceeds going to an appropriate charity).

Dismissal

10.46 The final power left to the employer is to dismiss the employee. This may be done for a number of reasons.

10.47 First, the circumstances may be so serious that a dangerous situation was created which put the employee or others at risk of serious injury. There are some activities where the degree of safety required is so high, and the consequences of a failure to achieve those standards so potentially serious, that a single departure from them could warrant instant dismissal. For example, in *Taylor v Alidair Ltd* (see above) it would be totally unrealistic to give the pilot a final warning saying 'If you land your aeroplane in such an incompetent manner again we will dismiss you. . .' etc. The driver of an express train, the scientist in charge of a nuclear power station, the driver of a vehicle carrying a dangerous chemical, etc, must all display the highest standards of care. In *Wilcox v HGS*, the employee was employed as a converter, changing gas appliances from town gas to natural gas. He was instructed that before he left premises he had to undertake a mandatory safety check, but failed to do so, and was dismissed. The dismissal was held to be fair by the industrial tribunal, although on appeal the case was remitted for reconsideration. If, as he alleged, the employers had persistently ignored the safety regulations, then such acquiescence was a relevant factor to be taken into account in determining whether or not a final warning should have been given.

10.48 Second, an employee may be in serious risk of injury to himself. In *Finch v Betabake (Anglia) Ltd* an apprentice motor mechanic was found to have defective eyesight. A report from an ophthalmic surgeon stated that the lad could not be employed without undue danger to himself and to others, and he was therefore dismissed. This was held to be fair. The fact that the employee is willing to take a risk that he may be injured is irrelevant, for the employer may expose himself to civil or criminal liabilities by continuing the employment (*Marsh v Judge International*).

10.49 Third, the situation may arise where the employee's physical condition is such that it amounts to a health or safety hazard. This must be handled carefully; there must be a full investigation, perferably backed with medical reports, there should be consultation with the employee, and alternative employment should be considered. In *Spalding v Port of London Authority*, the employee failed a medical examination after it was discovered that he was suffering from deafness. It was recommended that he worked with a hearing aid, but this did not prove to be satisfactory, and he was dismissed. This was held to be fair; the company's medical standards were not unnecessarily high, and were justified in order to ensure the safety of the employee and his fellow employees. In *Yarrow v QIS Ltd* the employee was dismissed after it was dicovered that he was suffering from psorisis. Because he had to work with radiography equipment, he was subject to the Ionising Radiation (Unsealed Sources) Regulations, which made it unsafe for him to be employed. This dismissal too was held to be fair. The employers were in danger of breaking the law if they continued to employ him. And in *Parsons v Fisons Ltd* the company's medical adviser noted that the employee had poor vision, and only narrowly averted several possible accidents. After a full discussion with the group medical adviser and her general practitioner, she was dismissed. This too, was held to be fair. There was no other suitable job for her, and it was not necessary to wait until an accident occurred before taking appropriate action.

10.50 It is important to bear in mind the health and safety of other employees, as well as their general comfort and working environment. In *Kenna v Stewart Plastics Ltd* the employee had a series of epileptic fits in an open plan office. The dismissal was held to be fair. The employer had a duty to ensure that other employees could do their work in reasonable working conditions which were physically and mentally conducive to work.

10.51 Fourth, the employee may be dismissed if he refuses to observe the safety instructions or wear the appropriate safety equipment. In *Frizzell v Flanders* the employee was provided with a gas mask while working in a tank. He was seen working without the mask and was dismissed. This was held to be fair. It was essential to enforce rigorously the safety precautions, and both for his own sake and for the sake of others, and his dismissal would serve as a warning that flagrant breaches of the safety instructions would not be tolerated.

10.52 An over-zealousness on the part of the employee to be cossetted against the risks of the employment can also result in his fair dismissal. In *Wood v Brita-Finish Ltd* the employee had been provided with acid-proof gloves, goggles, wellingtons and a protective apron, all of which had been approved by the Factory Inspectorate. He refused to work unless he was also provided with an overall, and was dismissed. This was held to be fair. Overalls had proved to be ineffective in the past, and contributed nothing to the safety of the employee. In *Howard v Overdale Engineering Ltd* the employee refused to work in a new factory because of dust caused by engineers drilling cables into the floor. The industrial tribunal found that the employers were not in breach of any statutory obligation to prevent impurities from getting into the air or to prevent employees from being subjected to harmful substances, and his dismissal was held to be fair for refusing to obey a lawful order.

10.53 It is not the function of the industrial tribunal to determine whether or not the employer is in breach of his common law or statutory duties, as they would not always have sufficient evidence available on which to make such a finding. In *Lindsay v Dunlop Ltd*, workers in the tyre-curing department became concerned about the possible carcinogenic nature of fumes and dust. As a temporary measure, it was agreed to resume normal working with masks being provided. The applicant, however, refused to adopt this course, maintaining that his continued exposure to fumes would endanger his health. His subsequent dismissal was held to be fair. Whether or not the employers were in breach of their obligations under section 63 of the Factories Act was a matter which could only be determined by the courts, and in the circumstances, the employers had not acted unreasonably.

10.54 To establish that it is fair to dismiss an employee for refusing to follow the safety rules, or to wear or use the precautions provided, it must be shown (a) the employee knew of the requirement, (b) the employer was consistent in his enforcement policies, and (c) the precautions were suitable for the employee and for the work he was doing. Again, a full investigation into the circumstances is called for. In *Mayhew v Anderson (Stoke Newington) Ltd* an insurance company recommended that the employee be asked to wear protective glasses, stating that the company's insurance cover would be withdrawn if she did not do so. The employers purchased a pair of safety goggles for 78p put she refused to wear them because they were not comfortable. She was warned that if she persisted with her refusal she would be dismissed, and ultimately the threat was carried out. Her dismissal was held to be unfair. She had never refused to wear reasonable eye protectors, only this particular type, which irritated her eyes and were uncomfortable. Custom-made eye protectors were available at a cost of £33, and the industrial tribunal thought that these should have been provided for her, even at the risk of creating a precedent.

10.55 Finally, other reasons prompted by a genuine concern for health and safety can justify dismissal. In *Wilson v Stephen Carter Ltd* the applicant was dismissed after refusing to go on a training course which involved staying away from home for a week, and this was held to be fair.

10.56 If lesser disciplinary sanctions do not succeed, the employer may ultimately dismiss a recalcitrant employee, but some employers do not consider this to be a satisfactory solution, as they would rather have the workers working, than have the problem of obtaining new staff and training them all over again. At this stage, it may be possible to invoke the assistance of the HSE Inspectorate, who could issue a Prohibition Notice on the employee, which would effectively prevent him from working in contravention of the matters contained in the Notice. A failure to comply with this is punishable by a fine or even imprisonment, and this may yet prove to be an effective way of dealing with the problem. At the same time, an employee could be warned that he is acting in breach of his duty under sections 7 or 8 of HSWA, which again is a criminal offence.

Suspension on medical grounds

10.57 Section 19 of the Employment Protection (Consolidation) Act provides that where an employee is suspended from work on medical grounds in consequence of

(a) any requirement imposed by or under the provision of any enactment, or
(b) any recommendation contained in a Code of Practice issued under HSWA,

which, in either case, is a provision specified in Schedule 2 of HSWA, then that employee shall be entitled to be paid during the suspension for a period of up to twenty-six weeks. The present provisions which are contained in Schedule 2 are as follows:

(1) Indiarubber Regulations 1922;
(2) Chemical Works Regulations 1922;
(3) Electric Accumulator Regulations 1925;
(4) Ionising Radiations (Unsealed Radio Active Substances) Regulations 1968;
(5) Ionising Radiation (Sealed Sources) Regulations 1969;
(6) Radioactive Substances (Road Transport Workers) Regulations 1970 and 1975;
(7) Control of Lead at Work Regulations 1980.

10.58 It will be noted that there must be a suspension on medical grounds. This means the potential effect on the health of the employee, not the actual effect. In other words, the provisions of section 19 are not relevant if an employee is actually off work sick. Nor is the suspension on medical grounds if he is unable to work because a Prohibition Notice has been imposed. Medical suspension payments can only be claimed when there is a suspension from work in order to comply with a requirement in any of the above provisions, but if he is incapable of working because of any physical or mental disablement, he has no legal entitlement (section 20(1)). Further, if he is dismissed because of one of the above requirements, he can bring a claim for unfair dismissal as long as he has been employed for a period of four weeks, instead of the more usual period of one year (two years in respect of employers who employ less than twenty employees). Whether such dismissal would be fair will obviously depend on the circumstances. For example, it may be shown that the employee's job became redundant, etc.

10.59 If the employer still needs someone to do the work, he may decide to take on a temporary replacement. He should inform the latter in writing that his employment will be terminated at the end of the period of suspension. If, therefore, he has to dismiss the temporary employee in order to permit the first employee to return to work, the dismissal will be for 'some other substantial reason', but without prejudice to the rule that the employer will still have to show that he acted reasonably in treating that reason as a sufficient ground for dismissal (EPCA, section 61). However, since the medical suspension period is unlikely to last long enough to enable the temporary employee to obtain a sufficient qualifying period of employment, this provision is somewhat otiose.

10.60 To qualify for medical suspension payments, the employee must have been employed for more than four weeks. Further, he will not be entitled to be paid if the employer has offered him suitable alternative work (whether or not the employee was contractually obliged to do that type of work) and he unreasonably refuses to perform that work. The employee must also comply with reasonable requirements imposed by the employer with a view to ensuring that his services are available. In other words, as the employer is paying the employee wages during the suspension, the employer may require the employee to do other work, or hold himself in readiness for work. The amount of pay to be made is calculated in accordance with Schedule 4 of EPCA, which depends on the contractual arrangements for pay.

10.61 An employee may complain to an industrial tribunal that the employer has failed to pay him in accordance with the above provisions. The complaint must be presented within three months of the failure, and if the complaint is upheld, the industrial tribunal will order the employer to pay the amount due.

10.62 An employee who does not work for more than sixteen hours per week (unless he works between eight and sixteen hours per week, and has done so for more than five years), or who works on a fixed term contract for ten weeks or less, is not covered by the above provisions (EPCA, section 143(3)).

11 The impact of international obligations

European Communities

11.1 In 1951, by the Treaty of Paris, the European Coal and Steel Community (ECSC) was established, when six countries (France, West Germany, Italy, Belgium, Holland and Luxembourg) agreed to pool their coal and steel resources and create a common commercial market for their products. Subsequently, the Mines Safety and Health Commission was created to work for the elimination of occupational risks to health and safety in coalmines.

11.2 In 1957, the European Atomic Energy Commission (EURATOM) was established in order to co-ordinate and develop the peaceful uses of nuclear energy, and strong emphasis was placed on the need to ensure the protection of the health of workers as well as the community at large from dangers arising from radiation hazards.

11.3 Also in 1957, by the Treaty of Rome, The European Economic Community (EEC) was established. This has the wider objective of establishing a common market for its economic activities by the elimination of customs duties, creating common customs tariffs, permitting the free movement of capital and workers, laying down common agricultural and transport policies, and harmonising the laws of Member States to ensure that competition is not distorted and to facilitate the anticipated economic expansion.

11.4 In 1967 these three institutions were merged into the European Communities (EC), with the fusion of their executive institutions, and the result is that although the three organisations have a separate existence, they are all now under the one umbrella.

By the European Communities Act 1972, the United Kingdom signified its accession to the Treaty of Rome, including its laws, which by section 2(1) of the Act, are to be given legal effect without further enactment. Ireland and Denmark joined at the same time as the United Kingdom, and Greece became a member in 1981.

The institutions of the European Communities

A COUNCIL OF MINISTERS

11.5 This is the final decision-making body on all major policy matters. It consists of the Foreign Ministers of Member States (although on occasions other Ministers will attend instead, e g the Ministers of Agriculture). Each Member State has the right of veto on proposals which affect its vital interests, but other votes are taken on the weighted majority principle, which recognises the relative unequal economic and political strengths of the respective countries.

B EUROPEAN PARLIAMENT

11.6 This consists of members elected by direct elections in the Member States. Each country has a number of seats based on its population, e g 81 MEPs are elected by France, Germany, Italy and Great Britain, with proportionately fewer from the smaller countries. The European Parliament is a public forum for the discussion of the political interests, and has the right to be consulted on proposals for new legislation. It can express its opinions on proposals which have been sent from the Commission to the Council of Ministers, can submit questions to both these institutions, and can, as a final sanction, dismiss the Commission.

C EUROPEAN COMMISSION

11.7 This may be regarded as a somewhat powerful civil service of the EC. It is controlled by thirteen Commissioners, nominated by the Member States, but not subject to any control by them. In particular, it has the right to initiate proposals for new legislation, will seek to ensure that Member States are conforming with European legislation, and can take enforcement action by referring alleged breaches of Community law to the European Court of Justice (for example, the Commission took the United Kingdom to the Court for its failure to introduce legislation enforcing the use of tachographs as required by an EC Directive, and the UK Government gave an undertaking to do so by 1981).

11.8 The Commission operates through twenty Directorates-General (DGs) with the addition of a number of other departments. In the main, Directorate-General V, which deals with Employment and Social Affairs, has an overall responsibility for health and safety matters, although there is an overlap with other DGs where matters of common concern arise. Directorate E of DG V is the department most closely concerned with health and safety mattres, and is based in Luxembourg. It has responsibilities for (a) toxicology, biology and health effects, (b) radioactive waste, accident prevention and safety measures in nuclear installations, (c) public health and radiation protection, (d) industrial medicine and hygiene, (e) industrial safety, and (f) Mines Safety and Health Commission.

11.9 In order to advise the Commission generally on all aspects of health and safety at work, an Advisory Committee on Safety, Hygiene and Health Protection was established in 1974. This consists of two members from governments, two from employers' associations, and two from trade unions, making a total of six from each Member State. This Committee assists the Commission in the preparation and implementation of activities in the field of safety, health and hygiene relating to work activities, with the exception of those areas which are dealt with by the Mines Health and Safety Commission and EURATOM.

11.10 The Advisory Committee's terms of reference are as follows:

(a) conducting, on the basis of information available to it, exchanges of views and experience regarding existing or planned regulations;
(b) contributing towards the development of a common approach to problems which exist in the field of safety, hygiene and health protection at work, and towards the choice of Community priorities, as well as the measures necessary for implementing those priorities;
(c) drawing the Commission's attention to areas in which there is an apparent need for the acquisition of new knowledge and for the implementation of appropriate educational and research projects;
(d) defining, within the framework of Community action programmes, and in co-operation with the Mines Safety and Health Commission, the criteria and aims of the campaign against the risks of accidents at work and health hazards within the undertaking, and methods enabling undertakings and their employees to evaluate and to improve the level of protection.

11.11 The Advisory Committee has been instrumental in assisting the Commission to draw up the fourteen point Action Programme (see below) on Safety and Health at work, which has been approved by the Council of Ministers.

D EUROPEAN COURT OF JUSTICE

11.12 This Court consists of ten judges appointed from the Member States to act as the final determining body on questions as to whether or not the Treaties setting up the three Community institutions are being correctly interpreted and applied by the Member States. Where there is a conflict between national law and Community law, the latter must prevail. A reference to the Court may be made by the Commission in respect of alleged violations by Member States, but although there is no machinery for the enforcement of the Court's decisions, a refusal to accept a ruling would strike at the very foundations of the European Community. Any court in this country may refer an issue to the European Court direct for a ruling on the meaning or application of the Treaty of Rome or any other European legislation, and in fact a number of questions have been so referred on equal pay claims (see e g *Macarthys Ltd v Smith*). To date, there has been no direct reference on any matter concerning health and safety at work.

European Community law

11.13 The prime source of European law is the various treaties which created the three Community institutions, together with the Merger Treaty, and which are directly applicable in Member States without any further national legislative action. Additionally there are the following legal instruments:

A REGULATIONS

11.14 These are of general application, binding in their entirety, and directly applicable in Member States. They must be introduced into national laws in the form in which they are expressed. For technical reasons, it is not intended that Regulations shall be made relating to health and safety matters at this stage.

B DIRECTIVES

11.15 These are binding instructions to Member States as to the results which must be achieved by national legislation, but the Member States have a discretion as to the means of implementation. Sometimes, Framework Directives are issued on a number of related subjects, stating overall principles and objectives. The Commission will issue Directives containing technical details, and will frequently issue amending Directives in accordance with changing technology. Although Directives are addressed to nation States, and are thus not binding on their subjects, it has been held by the European Court that if national law conflicts with the provisions of a Directive, an individual may plead the Directive in litigation, and the national court must resolve the issue as if the nation State had enacted the Directive (*Van Duyn v Home Office*). However, it is thought that only certain types of Directives are directly applicable in this manner.

11.16 Nearly all Directives relating to health and safety are made under the provisions of article 100 of the Treaty of Rome, which is concerned with the removal of barriers to trade. Since health and safety measures can be expensive, it would amount to unfair competition if employers in one country complied with high standards, whilst those in another country did not, and were thus able to undercut their international competitors. A further motive for the improvement of international standards is to benefit workers generally throughout the EC, as well as to facilitate the free movement of workers and capital.

C DECISIONS

11.17 These are rulings given by the Commission in individual cases, and may be addressed to a Member State, or to a company, or to an individual. A Decision may be used as a means of giving permission to a State to depart from Community law, and is thus binding on the person to whom it is addressed, but applies to no-one else.

D RECOMMENDATIONS AND OPINIONS

11.18 These have no binding force whatsoever, but are mere statements as to the views of the Commission on certain subjects, and are used as a means of encouraging uniformity of practice throughout the Community.

Consultative bodies

11.19 Before EC legislation is passed, a tremendous amount of consultative work takes place. Proposals need to be supported by relevant scientific or technical data or surveys, national experts are consulted, advisory committees are asked for opinions, and initial proposals will then be drawn up. These are then transmitted to the European Parliament, to the Economic and Social Committee, and to the Council of Ministers. At all stages, representations can be made by national groups representing employers, trade unions and other interested parties.

11.20 When proposals reach the United Kingdom, a Government Department (generally known as the 'lead Department') which is most closely concerned with the proposal, will take charge of the consultative process. Usually, on health and safety matters, this will be the Department of Employment, but it may be some other Department which has an major interest in the proposals (e g Department of the Environment). Discussions will continue with the TUC, CBI, trade associations, etc, and an explanatory memorandum will be prepared by the lead Department and submitted to Parliament for consideration by the scrutiny committee of each House. These committees may call for written or oral evidence, make recommendations for change, or request a Parliamentary debate. Once the Government has formulated its views, the matter can be transmitted back to the Council of Ministers for consideration by a working group. Here, the respective views are collated, the text may be revised, and the final proposals formulated and ultimately adopted. Not surprisingly, it can take many years before an initial proposal is finally transformed into a binding Directive.

11.21 Once a Directive relating to health and safety at work has been adopted, HSC draws up the necessary legislative proposals in order to implement it. Again, it will engage in a series of consultations with interested parties, but since all concerned should have been involved in the earlier discussion, the subject matter will occasion little surprise, and the only problems which are likely to arise will stem from the detailed arrangements which may be necessary in order to ensure that the final legislative proposals (usually made by Regulations) will meet the European standards. Indeed, HSC takes pride in considering that it already anticipates European legislation as part of its own ongoing programme, and is thus in a favourable position to influence the European standards.

EC Action Programme

11.22 In June 1978, the Council of Ministers approved a fourteen point Action Programme to be pursued by the Commission, to be completed by 1982. The programme was designed to implement a policy of prevention of risks, with intense controls where risks cannot be eliminated. Three broad objectives were thus identified as guiding principles for the Programme. The first was the need to expand the co-operation between Member States with regard to measures taken concerning safety, health and hygiene at work. As well as harmonising legislation, Member States should inform the Commission of draft proposal for new laws, Codes of Practice, etc, so as to enable an exchange of ideas to take place. There could also be co-ordination of knowledge relating to the identification of hazards at work, and the steps taken to eliminate or reduce them. The methodology of statistical analysis could be improved or standardised, and a data bank of research projects created, so as to avoid duplication of effort and to facilitate collaboration. The second objective would be to take steps necessary to bring about improvements within the organisation of work. This would involve preventative and monitoring measures, integrating health and safety into the design of premises, plant and machinery and into the manufacturing stages. Of particular importance would be the exchange of information concerning major hazards which affect the Community at large, so that the lessons of Flixborough and Seveso should not be repeated elsewhere. Third, steps must be taken to bring about a change in human attitudes, so as to develop and strengthen safety consciousness throughout the Community. Safety education can start at a fairly young age, and be pursued throughout the working life. At the same time, management and industrialists generally can stimulate safety by taking active interest in the subject.

11.23 The Commission is required to produce an annual report for the Council of Ministers on the progress achieved in pursuing the Action Programme. The Advisory Committee on Safety, Hygiene and Health Protection is closely associated with the Commission's work, and there is liaison with the European Foundation for the Improvement of Living and Working Conditions. The following is a summary of the Action programme, with a resume of the progress made during the first year (taken from the Commission's first report made in February 1980).

(1) To establish a common statistical methodology in order to assess with sufficient accuracy the frequency, gravity and causes of accidents at work, and also the mortality, sickness and absenteeism rates in the case of diseases connected with work.

Progress. The Community has statistics on accidents in the coal and steel industry, and studies are under way in the construction industry. However, due to lack of manpower, further developments on this programme have been limited.

(2) To promote the exchange of knowledge, establish close co-operation between research institutes, and identify subjects suitable for joint research.

Progress. Research programmes in the coal and steel industry are continuing, and will be used as a model for other sectors. A framework research programme will be drawn up in 1981, and a feasibility study has been undertaken concerned with the documentation system for all research projects in progress in the field of occupational health.

(3) The standardisation of terminology and concepts relating to exposure limits for toxic substances, and to harmonise exposure limits, taking into account those which are already in existence.

Progress. A study has been completed entitled 'Comparative Analysis of the Principles and Application of Control Limits in Member States of the European Community'. This shows the wide differences in terminology which currently exist. Meetings of national experts are being held with a view to providing a basis for the harmonisation of concepts and standards. The International Labour Organisation and the World Health Authority are co-operating with this programme.

(4) To develop a preventative and protective action against recognised carcinogenic substances by fixing exposure limits, sampling requirements, and measuring method and satisfactory conditions of hygiene at work, specifying prohibitions where necessary.

Progress. A Directive dealing with exposure to vinyl cloride monomer has already been issued by the Council, and collaborative work is continuing with the International Agency for Research on Cancer. A general Directive dealing with the principles of prevention and protection is currently under discussion.

(5) In respect of certain specified toxic substances (asbestos, arsenic, cadmium, lead and chlorinated solvents) to establish exposure limits, threshold limit values, sampling requirements and measuring methods, and satisfactory conditions of hygiene at the workplace.

Progress. A proposed Directive has been sent to the Council dealing with the protection of workers from harmful exposure to chemical, physical and biological agents at work, which will require Member States to enact framework legislation which will ensure that exposure to harmful substances will be kept as low as can reasonably be achieved. This may be done by a number of measures, including introducing prevention at the technical level, establishing exposure limits, introducing protection measures, setting hygiene requirements, providing information to workers on the risks involved, and the preventative measures to be taken, setting requirements for employers to follow, introducing warnings and safety signs, medical surveillance, keeping necessary records, and introducing emergency procedures for abnormal exposures. An Industrial Medicine and Hygiene Committee has been set up to deal with the technical and scientific investigations which are needed in this field. The Commission has also submitted a proposal for a Directive on lead, and a further one dealing with asbestos is in the process of being drawn up.

(6) To establish a common methodology for the assessment of health risks connected with the physical, chemical and biological agents present at the workplace, in particular by research into the criteria of harmfulness and by determining the reference values from which to obtain exposure limits.

Progress. A number of studies dealing with dose/effect relationship have been published, which are complemented by work undertaken by other international organisations (ILO, WHO etc). The results are used by the Commission in evaluating the risks and for drawing up of proposed Directives.

(7) To establish information notices and handbooks on the risks relating to the handling of a certain number of dangerous substances, such as pesticides, herbicides, carcinogenic substances, asbestos, arsenic, mercury, lead, cadmium and chlorinated solvents.

Progress. The Council Directive on Safety Signs at Work is an important step in providing a common basis for information for workers, concerning dangers at the workplace and the necessary precautions to be taken. Further studies are being made into international methods regarding signs.

(8) To establish the limit levels for noise and vibrations at the workplace,, and determine practical ways and means of protecting workers and reducing sound levels at the place of work. Also, it is intended to establish the permissible sound levels of building site equipment and other machines.

Progress. A report entitled 'Damage and Annoyance caused by Noise' has already been published. The Commission will use the report as background information in the drawing up of proposals.

(9) To undertake a joint study of the application of the principles of accident prevention and of ergonomics in the design, construction and utilisation of plant and machinery, and promote the application of these principles in certain pilot areas, including agriculture.

Progress. The majority of the work in this area has been carried out in the coal and steel industry. The Commission is currently studying how similar principles can be applied to the construction industry and to agriculture.

(10) To analyse the provisions and measures governing the monitoring of the effectiveness of safety and protective arrangements, and to organise an exchange of experience in this field.

Progress. Studies have been instituted concerned with toxico-vigilence systems, designed to give an early warning of new and previously unknown hazards. A workshop has been held in conjunction with the ILO on a hazard alert system, with a view to passing on information about newly discovered hazards and protection methods, and the use of an international network of collaborating centres for this purpose.

(11) To develop a common methodology for the monitoring of pollutant concentrations and the measurement of environmental conditions at the place of work, to carry out inter-comparison

programmes and establish reference methods for determining the most important pollutant, to promote new monitoring and measuring methods for the assessment of individual exposure, in particular through the application of sensitive biological indicators. Special attention is to be given to the monitoring of exposure in the case of women, with special reference to pregnant women and to adolescents. To undertake a joint study of the principles and methods of the application of industrial medicine with a view to promoting better the protection of workers' health.

Progress. The Commission has already carried out work on the quality assurance of lead measurements, and intends to adapt this to work situations. A series of monographs will be published on biological monitoring, with update reviews of available data. A further study has been made of occupational medical services, and proposals are expected to emerge shortly.

(12) To establish the principles and criteria applicable to the special monitoring relating to assistance or rescue teams in the event of accident or disaster, and to maintenance and repair teams, and the isolated worker.

Progress. The proposed Directive on the risk of major accidents in certain industrial activities is aimed at the prevention of such incidents and reducing their adverse consequences. If the industrial activity is likely to lead to serious consequences if an accident occurs, a safety report must be drawn up, workers informed, equipped and trained, safety drills organised, and the neighbouring population informed and an emergency plan established. Member States are to inform the Commission when such an accident occurs, and the Commission will keep a data bank with the information stored therein.

(13) To exchange experience concerning the principles and methods of inspection by public authorities in the field of safety, hygiene at work, and occupational medicine.

Progress. The Commission has planned a joint study together with the ILO and the European Foundation for the Improvement of Living and Working Conditions.

(14) To draw up induction and information schemes on safety and hygiene in respect of particular categories of workers, especially migrant workers, newly recruited workers, and workers who change their jobs.

Progress. Because of manpower shortages, no action has been taken on this programme.

Council Directives

11.24 The following is a list of Directives relating to health and safety which have been adopted by the Council of Ministers, with the year of adoption stated.

(1) Electrical Equipment for use in Potentially Explosive Atmospheres (Framework Directive) 1976;
(2) Pressure Vessels (Framework Directive) 1976;
(3) Classification, Packaging and Labelling of Dangerous Preparations (Solvents) 1973;
(4) Classification, Packaging of Paints, Varnishes and Glues 1977;
(5) Wire Ropes, Chains and Hooks 1973;
(6) Safety Signs at Work 1977;
(7) Euratom: Basic Safety Standards for Radiological Production 1976;
(8) Toxic Products 1974;
(9) Noise: Method of Measurement 1979;
(10) Protection of Health of Workers exposed to Vinyl Chloride 1978;
(11) Driver-pierced Noise Level (Agricultural Tractors) 1977;
(12) Roll-over Protective Structures (Agricultural) 1977;
(13) Passenger Seats (Agricultural Tractors) 1976;
(14) Classification, Packaging and Labelling of Dangerous Preparations (Pesticides) 1978;
(15) Protection of Workers from the risks related to exposure to Chemical, Physical and Biological agents at Work 1980;
(16) Classification, Packaging and Labelling of Dangerous Substances 1967 (as amended by six further Directives).

International Labour Organisation (ILO)

11.25 The ILO was formed in 1919 and consists of representatives of national governments, employers and workers organisations. It has worked consistently to improve international labour standards relating to such matters as conditions of work, training, freedom of association, social security, industrial relations, and many other similar topics. It holds international conferences, provides technical advice and assistance to individual countries, and generally acts as an international forum for the promotion and improvement of standards throughout the world.

11.26 A major part of the work of the ILO consists of adopting Conventions and Recommendations. These are submitted to national governments for consideration, for they are not automatically binding. A Convention may be ratified by a nation state, which amounts to a pledge to implement its provisions. A Recommendation does not require ratification, but merely serves as a guide if national action is to be taken on a particular topic.

11.27 Since its inception, the ILO has passed over 150 Conventions, many relating to occupational health and safety matters. It has also passed over 160 Recommendations covering similar topics, and has published a large number of reports and studies.

11.28 The ILO also produces research papers, suggests international classification standards, and issues Codes of Practice giving guidance on practical measures which may be taken to safeguard workers' health against occupational hazards.

12 Future proposals

Notification of new substances

12.1 The continuous search for new substances has brought about a dramatic increase in synthetic organic chemicals, which, in turn, has caused growing concern over the number of people who are exposed to their acute or long-term effects, as well as having environmental and pollution implications. In 1979 the Council of the European Communities adopted a Directive (79/831) requiring each member country to introduce a scheme for the testing and notification of all new substances which are placed on the market and the draft Notification of New Substances Regulations are currently being considered.

12.2 The basis of the scheme is that manufacturers and importers must test all new substances they intend to market in quantities of one tonne or more per annum (less than one tonne if the substance is toxic). At least forty-five days before the substance is placed on the market HSE must be informed, and given certain data relating to the scale of manufacture, intended usage and the properties of the substance. The notifier must submit a dossier containing details of the identity of the substance, its physical, chemical and biological properties, and a statement of the precautions which must be observed for its safe handling and use, with further information concerning its appropriate labelling. The biological tests must be designed to determine the acute effects (caused by oral, dermal or inhalation contact) the chronic effects (caused over a period of time) and the special effects (for carcinogenity, mutagenecity and teratogenicity). HSE can then determine if further testing is necessary, and require this to be done, with an appeal procedure if the notifier feels that this is not reasonably necessary.

12.3 Three Approved Codes of Practice will accompany the new regulations. These are:

(a) Testing for potential effects on human health and the environment;
(b) Good laboratory practice;
(c) Tests for physico-chemical properties.

12.4 Current schemes (e g those controlling the marketing of pesticides and new drugs) will come within the new Regulations. There is no power to actually prevent the marketing or sale of any new substance, only to require further testing. However, it is not thought this is omission will be significant, as those products which are marketed will be required to have a full toxicological profile.

12.5 The new scheme will also complement the existing obligations contained in section 6 of HSWA (to test substances, see chapter 3), but there is the additional requirement to submit the results of such testing to HSE for scrutiny, and HSE could demand further data if there was any doubt about the safety of the new substances. Confidentiality, which is essential for the protection of commercial interests, will be preserved (see HSWA, section 28). If production exceeds 100 tonne per year, or when it exceeds 1,000 tonne per year, HSE must be further notified.

Homeworkers

12.6 A homeworker is a person who

(a) works in domestic premises, and
(b) is provided with either
  (i) materials on which to work, or
  (ii) instructions on which materials he should purchase in order to do the work, and
(c) makes available the product of the work to the person providing the materials or instructions, or to another designated person.

12.7 This definition distinguishes the homeworker from self-employed craftsmen who are responsible for marketing their own produce, whereas a homeworker contributes to the products which are marketed by his employer. The self-employed person is responsible for his own health and safety, whereas the person responsible for putting out work to the homeworker bears the prime responsibility for ensuring that, so far as reasonably practicable, no risks to health and safety arise during the course of the work.

12.8 If the homeworker is an employee, his employer is bound by the duties in section 2 of HSWA (see chapter 3). In any case, the employer is bound by section 3 (duties to non-employees) and manufacturers etc of articles and substances to be used at work are bound by section 6 in relation to homeworkers as they are to anyone else.

12.9 The Proposed Health and Safety (Homework) Regulations will apply to all homework other than work normally performed in office premises or where the product of the work is exclusively for the personal or domestic use of the person providing the materials or instructions. In other words, work which is of a clerical nature (e g typing addressing envelopes etc) is excluded, as is a private domestic laundry service but if the person providing the material or instruction requires the homeworker to use potentially hazardous processes then obviously the general requirements of sections 2–3 of HSWA apply.

12.10 Every person who provides homework shall send to the enforcing authority twice each year (in February and August) certain information, namely:

(a) the name and address of the person providing the material for homework;
(b) the address of the premises from which the material for homework is provided;
(c) the nature of the homework;
(d) a full and proper description of any material or equipment provided to the homeworker or which he is instructed to purchase;
(e) the number of homeworkers provided with homework within the preceding six months;
(f) the name and address of any third party through whom any homeworker is in contact with the person who provides the homework;
(g) if the enforcing authority makes a written request, the names and addresses of all or any specified group of homeworkers for whom work is provided.

12.11 Additionally, a list must be kept available of the full names and addresses of all persons to whom material for homework has been provided during the three years immediately preceding the date on which the above information was last sent to the enforcing authority.

12.12 Certain substances may not be provided for use in homework without prior consent being given by the enforcing authority. These include:

(a) certain radioactive substances;
(b) asbestos (including various compounds);
(c) mercury (unless permanently sealed in equipment);
(d) most aromatic amines and their salts;
(e) silica;
(f) highly flammable liquids;
(g) compounds of lead;
(h) isocyanates.

12.13 Any consent which has been given may be made subject to conditions, may be limited in duration and may be revoked at any time. Section 133 of the Factories Act, and the various Home Work Orders will be repealed.

Plan of work

12.14 In June 1981 HSC published its 'Plan of Work 1981–82 and onward', which is a review of all the proposed activities, together with background information. Also published was a list of factors which would be taken into account in determining the content of the work programme and any priorities of order. As the Chairman of HSC has pointed out, items in the work programme which involve the preparation of Regulations will inevitably attract particular attention, but these represent only a small part of the activities of HSC. Most of the staff and resources will be devoted to advisory work and seeking compliance with existing legislation, to assessment work, and to health and safety research.

12.15 In all, forty-nine projects for work on Regulations and Approved Codes of Practice were listed, of which seven arise from the implementation of EC Directives, seven relate to mining and quarrying industries, and ten relate to standards for use by British industry in agreement with the Department of Industry.

12.16 *Future proposals*

12.16 The Plan of Work has been submitted to the Secretary of State, which he may approve, disapprove, or modify. Subject to this, the Plan of Work will consider the following matters:

i *Major hazards*

12.17 The European Commission is at present engaged in discussions concerning the draft Directive on Major Accident Hazards of Certain Industrial Activities (the 'Seveso' Directive) a project in which HSC is obviously taking a great interest. However, HSC may go ahead with its own plans for the preparation of the Hazardous Installations (Notifications and Survey) Regulations, which will anticipate in form and structure the EC Directive.

ii *Explosives*

12.18 Much of the existing legislation on this topic is out of date, and there is need to harmonise with international developments. Among the topics to be considered are new Regulations on the classification and labelling of explosives, improving security controls, Regulations on conveyance by road of explosives, and a revision of Regulations relating to the use of explosives in mines and quarries.

iii *Dangerous substances*

12.19 Regulations on this topic will need to be constantly revised in order to implement EC Directives. A new single set of Regulations, Approved Code of Practice and Guidance Notes will cover classification, packaging and labelling of dangerous substances, including preparations, for supply and conveyance by road.

iv *Conveyance of dangerous substances*

12.20 A coherent set of controls will be produced covering all aspects of the conveyance of dangerous substances by all modes of transport, including tankers and tank containers, packaged goods, and the control of dangerous goods in ports and harbours.

v *Toxic substances*

12.21 As already noted (para 12.1) progress is being made on the proposed Notification of New Substances Regulations. Further work is being done on the control of substances which are hazardous to health, including asbestos and certain fumigants.

vi *Flammable substances*

12.22 Existing controls relating to flammable substances are in need of revision, and do not cover all work activities. Across the board Regulations will modernise the legal requirements.

vii *Noise*

12.23 The only legislative requirements which relate to noise concern the restrictive areas of woodworking, tractor cabs and offshore installations. New Regulations, an Approved Code of Practice and Guidance Notes will be prepared on the protection of hearing at all work activities (see below, para 12.36).

viii *Mechanical lifting systems*

12.24 New Regulations and an Approved Code of Practice on shafts and winding in mines are being considered, as the existing statutory requirements are fragmentary.

ix *Pressurised systems*

12.25 Existing fragmentary statutory requirements will be replaced by up-to-date Regulations and guidance. In particular the British Standards Institute has put forward a number of proposals for an Approved Code of Practice.

x *Transport*

12.26 Existing Regulations on the provision of safe systems of underground transport in mines do not take account of technological change, and as this area has a poor safety record, new Regulations and an Approved Code of Practice will be considered.

xi *Other machinery and equipment*

12.27 Existing statutory requirements will eventually have to be replaced, but it is conceded that this area has a low priority. However, certain proposals put forward by the British Standards Institute for Approved Codes of Practice in respect of certain technology will be considered.

xii *Ionising and non-ionising radiations*

12.28 New Regulations, Approved Codes of Practice and technical guidance will shortly be prepared with a view to implementing the Euratom Directive on Basic Standards of Radiological Protection and updating existing legislation.

xiii *Electricity*
12.29 Present Regulations only cover about one-third of the workforce, and use outmoded terminology. New Regulations will take into account modern developments, and will apply to all work situations. Certain proposals put forward by the British Standards Institute for Approved Codes of Practice will also be considered.

xiv *Mines and quarries*
12.30 New Regulations are proposed dealing with safety lamps and lighting, and ventilation in mines.

xv *Agriculture*
12.31 The Health and Safety (Agriculture) (Poisonous Substances) Regulations require amendment to take into account substances which are no longer used, and the introduction of new substances. Various proposals relating to agricultural tractors will also be considered.

xvi *Docks and offshore installations*
12.32 Regulations, Approved Codes of Practice and Guidance Notes will be produced on Safety Representatives and Safety Committees, and for first aid in respect of offshore installations.

xvii *Other projects*
12.33 A number of other minor projects are listed, including a review of the legislation which discriminates between men and women, increase in fees to be charged for various services provided by HSC and EMAS, allocating responsibility for the administration of industrial air pollution control, and giving approval to Codes of Practice in respect of certain proposals put forward by the British Standards Institute.

Metrication

12.34 For many years it has been the policy of successive governments to introduce metric standards into this country, and this is now required by an EC Directive. HSWA, section 49 enables the appropriate Minister to make Regulations to amend any of the relevant statutory provisions and certain other enactments and statutory instruments by substituting an amount or quantity expressed in metric units for an amount not so expressed. These new measurements may be 'rounded off' in order to express them in convenient and suitable terms (section 49(2)), but due consideration will always be given to the maintenance of safety standards. However, any plant, building or structure which were in existence before the changeover, and which complied with the former statutory imperial measures will not be affected.

12.35 This process of metrication has already started (e g see Docks, Shipbuilding etc (Metrication) Regulations 1981, and it is now proposed to further this by the Factories Act 1961 etc (Metrication) Regulations which, at the time of writing, have been produced in draft form. These provide that various provisions of the Factories Act which are expressed in imperial measurements shall have the metric equivalent substituted. Also being considered is the Agriculture (Metrication) Regulations which have been submitted to the Secretary of State for approval.

Occupational deafness

12.36 HSC has recently produced a Consultative Document entitled 'Protection of Hearing at Work' which outlines proposals for new Regulations, an Approved Code of Practice and Guidance Notes designed to reduce the risk of hearing loss due to noise at work. The Commission accepts that to impose an absolute requirement for all noise to be reduced to one limit would impose an intolerable burden on industry, and may lead to a complete closure in some areas. The proposals are designed to regulate those areas where the risk is greatest, and bring about a consequent reduction in the number of workers suffering from hearing loss.

12.37 The first general requirement will be for employers to ensure that no person shall be exposed above $90\text{dB(A)}L_{eq(8hr)}$ (i e an average of 90 decibels over an eight-hour day), or to an instantaneous sound pressure of 600 Pascals (Pa). To meet this requirement, the employer will be obliged to reduce noise levels to the lowest level reasonably practicable. If the reduction in noise levels does not reduce exposure to the required limits, the employer may achieve the desired results by other means (ear protectors, etc). If it is reasonably practicable to reduce noise levels to less than 90dB(A) or 600 Pa there will be a duty to do so.

12.38 Where exposure is likely to exceed these limits, disregarding any personal ear protection used, the employer will be under a duty to arrange for noise surveys, provide information, instruction and training, check control measures, produce an action programme, appoint a qualified noise adviser, and keep exposure records. If the exposure is likely to be above 105dB(A)$L_{eq(8hr)}$ employers will have to arrange audiometric testing to be carried out, and for monitoring of individual employees' exposure to noise.

12.39 Employees, for their part, will be under a duty to make full use of control measures, and to report any defect discovered in the equipment. They must also co-operate with the employer in the carrying out of the latter's duties, in particular in the use of equipment for surveys and individual monitoring of exposure to noise, and measures for control of noise exposure.

12.40 A further requirement will be placed on designers, manufacturers, importers and suppliers of articles for use at work. These persons will be required to ensure, so far as is reasonably practicable, that an article, when properly used, does not produce noise likely to be injurious to hearing. In determining proper use, no account is to be taken of the use of ear protectors. If the article is likely to exceed the limits of 90dB(A)$L_{eq(8hr)}$ or 600 Pa, adequate information must be made available about the noise levels, and measures which need to be taken to minimise the resulting exposure. There will also be a requirement to arrange for testing, examination and research.

12.41 HSE will be empowered to approve ear protectors, and to grant exemptions having regard to the circumstances of the case. It is also intended to revoke regulation 44 of the Woodworking Machine Regulations 1974.

Summary of the proposed Regulations

12.42

	Exposure likely to be		
	below 90dB(A) $L_{eq(8hr)}$ and 600 pascals	above 90dB(A) $L_{eq(8hr)}$ or 600 pascals	above 105dB(A) $L_{eq(8hr)}$
Duties of employers to reduce exposure likely to be injurious to hearing to the lowest level reasonably practicable	*	*	*
to reduce exposure by (a) reduction of noise levels to the lowest level reasonably practicable		*	*
(b) any other means		*	*
to arrange for surveys		*	*
to provide information, instruction and training		*	*
to provide ear protectors		*	*
to check that control methods etc are used		*	*
to produce a programme of action		*	*
to appoint a qualified person to advise		*	*
to keep records of exposure		*	*
to arrange for audiometry			*
to arrange for individual monitoring of exposure			*
Duties of employees to make full use of control measures	*	*	*

	Exposure likely to be		
	below 90dB(A) $L_{eq(8hr)}$ and 600 pascals	above 90dB(A) $L_{eq(8hr)}$ or 600 pascals	above 105dB(A) $L_{eq(8hr)}$
to co-operate with employer	*	*	*
Duties of designers, manufacturers, importers and suppliers of articles for use at work			
to ensure, so far as is reasonably practicable, that noise produced is not likely to be injurious to hearing	*	*	*
to carry out testing, examination and research		*	*
to ensure that information is available		*	*

Appendices

CONTENTS

Appendix A
Addresses of HSC, HSE and EMAS Headquarters and area offices

Health and Safety Commission

Regina House,
259/269 Old Marylebone Road,
London NW1 5RR,
Tel: 01-723 1262

Health and Safety Executive

25 Chapel Street,
London NW1 5DT,
Tel: 01-262 3277

Public Enquiry Point

Baynards House,
1 Chepstow Place,
Westbourne Grove,
London W2 4TF,
Tel: 01-229 3456 ext 68

HSE area organisations

Area	Address	Telephone No.	Local authorities within each area	National Industry Group (NIG)
1 SOUTH WEST	Inter City House, Mitchell Lane, Bristol BS1 6AN	0272 290681	Avon, Cornwall, Devon, Gloucestershire, Somerset, Isles of Scilly	Hospitals
2 SOUTH	Priestley House, Priestley Road, Basingstoke RG24 9NW	0256 3181	Berkshire, Dorset, Hampshire, Isle of Wight, Wiltshire	Rubber and research establishments
3 SOUTH EAST	3 East Grinstead House, London Road, East Grinstead, West Sussex RH19 1RR	0342 26922	Kent, Surrey, East Sussex, West Sussex	Paper and board manufacture
4 LONDON NW	Chancel House, Neasden Lane, London NW10 2UD	01-459 8844	Barnet, Brent, Camden, City of London, Enfield, Hammersmith, Harrow, Hillingdon, Hounslow, Kensington & Chelsea, City of Westminster	Printing and bookbinding
5 LONDON NE	Maritime House, 1 Linton Road, Barking, Essex IG11 8HF	01-594 5522	Barking, Hackney, Haringey, Havering, Islington, Newham, Redbridge, Tower Hamlets, Waltham Forest	Docks

Area		Address	Telephone No.	Local authorities within each area	National Industry Group (NIG)
6	LONDON S	1 Long Lane, London SE1 4PG	01-407 8911	Bexley, Bromley, Croydon, Greenwich, Kingston-upon-Thames, Lambeth, Lewisham, Merton, Richmond-upon-Thames, Southwark, Sutton, Wandsworth	Construction
7	EAST ANGLIA	39 Baddow Road, Chelmsford, Essex CM2 0HL	0245 84661	Essex except the London Boroughs in Essex covered by Area 5; Norfolk, Suffolk	Food and packaging
8	NORTHERN HOME COUNTIES	14 Cardiff Road, Luton LU1 1PP	0582 34121	Bedfordshire, Buckinghamshire, Cambridgeshire, Hertfordshire	Woodworking and furniture
9	EAST MIDLANDS	Belgrave House, 1 Greyfriars, Northampton NN1 2LQ	0604 21233	Leicestershire, Northamptonshire, Oxfordshire, Warwickshire	Plastics, footwear and leather
10	WEST MIDLANDS	McLaren Bldg, 2 Masshouse Circus, Queensway, Birmingham B4 7NP	021 236 5080	West Midlands	Foundries and general engineering
11	WALES	Brunel House, 2 Fitzalan Road, Cardiff CF2 1Sh	0222 497777	Clywd, Dyfed, Gwent, Gwynedd, Mid Glamorgan, Powys, South Glamorgan, West Glamorgan	Steel
12	MARCHES	The Marches House, The Midway, Newcastle-under-Lyme, Staffs ST5 1DT	0782 610181	Hereford and Worcester, Salop, Staffordshire	Ceramics

Area	Address	Telephone No.	Local authorities within each area	National Industry Group (NIG)
13 NORTH MIDLANDS	Birkbeck House, Trinity Square, Nottingham NG1 4AU	0602 40712	Derbyshire, Lincolnshire, Nottingham	Electricity generation and distribution
14 SOUTH YORKSHIRE	Sovereign House, 40 Silver Street, Sheffield S1 2ES	0742 739081	Humberside, South Yorkshire	Wire, rope and cable-making
15 W & N YORKS	8 St Paul's Street, Leeds LS1 2LE	0532 446191	North Yorkshire, West Yorkshire	Wool textiles
16 GREATER MANCHESTER	Quay House, Quay Street, Manchester M3 3JB	061-831 7111	Greater Manchester	Bottom textiles (and allied fibres)
17 MERSEYSIDE	The Triad, Stanley Road, Bottle L20 3PG	051-922 7211	Cheshire, Merseyside	Chemicals
18 NORTH WEST	Victoria House, Ormskirk Road, Preston PR1 1HH	0772 59321	Cumbria, Lancashire	New entrants (except health, research and education)
19 NORTH EAST	Government Buildings, Kenton Bar, Newcastle-upon-Tyne NE1 2YX	0632 869811	Cleveland, Durham, Northumberland, Tyne & Wear	Ship-building and ship repairing
20 SCOTLAND EAST	Meadowbank House, 153 London Road, Edinburgh EH8 7AU	031-661 6171	Borders, Central, Fife, Grampian, Highland, Lothian, Tayside, and the island areas of Orkney & Shetland	Distilling, brewing and other drink manufacturing
21 SCOTLAND WEST	314 St Vincent Street, Glasgow G3 8XG	041-204 2646	Dumfries and Galloway, Strathclyde, and the Western Isles	Education

Employment Medical Advisory Service

25 Chapel Street,
London NW1 5DT,
Tel: 01-262 3277

REGIONAL MEDICAL ADVISERS: INQUIRY POINTS

Region	*Address and Telephone No*
NORTHERN	Wellbar House, Gallowgate, Newcastle-upon Tyne NE1 4TP, Tel: 0632 27575 ext 565
YORKSHIRE AND HUMBERSIDE	City House, Leeds LS1 4JH, Tel: 0532 38232 ext 268
EASTERN AND SOUTHERN	Bryan House, 76 Whitfield Street, London W1P 6AN Tel: 01-636 8616 ext 448
LONDON AND SOUTH EASTERN	Hanway House, Red Lion Square, London WC1R 4NH Tel: 01-405 8454 ext 325
SOUTH WESTERN	The Pithay, Bristol BS1 2NQ Tel: 0272 21071 ext 356
WALES OFFICE	Dominions House, Queen Street, Cardiff CF1 4NS Tel: 0222 32961 ext 266
MIDLANDS	Fiveways House, Islington Row, Birmingham B15 1SG Tel: 021-643 9868 ext 476
NORTH WESTERN	Sunley Buildings, Piccadilly Plaza, Manchester M60 7JS Tel: 061-832 9111 ext 2110
SCOTTISH HEADQUARTERS	Stuart House, 30 Semple Street, Edinburgh EH3 8YX Tel: 031-299 2433 ext 169

Appendix B
HSE Improvement and Prohibition Notices

Improvement Notice: Form LP1

HEALTH AND SAFETY EXECUTIVE **Serial No.I**
Health and Safety at Work etc Act 1974, Sections 21, 23 and 24

IMPROVEMENT NOTICE

Name and address (See Section 46)	To ....
	....
(a) Delete as necessary	(a) Trading as ....
(b) Inspector's full name	(b) ....
(c) Inspector's official designation	one of (c)..
(d) Official address	of (d) ....
	.... Tel No. ....
	hereby give you notice That I am of the opinion that at
(e) Location of premises of place and activity	(e)..
	you, as (a) an employer/a self employed person/ a person wholly or partly in control of the premises
(f) Other specified capacity	(f) ....
	(a) are contravening/have contravened in circumstances that make it likely that the contravention will continue or be repeated
	....
	....
(g) Provisions contravened	(g) ....
	....
	The reasons for my said opinion are:-...............
	....
	....
	and I hereby require you to remedy the said

	contraventions or, as the case may be, the matters occasioning them by
(h) Date	(h) ....
	(a) in the manner stated in the attached schedule which forms part of the notice.
	Signature .... Date................
	Being an inspector appointed by an instrument in writing made pursuant to Section 19 of the said Act and entitled to issue this notice.
	(a) An improvement notice is also being served on
	....
	of ....
LP1	related to the matters contained in this notice.

Prohibition Notice: Form LP2

HEALTH AND SAFETY EXECUTIVE

Health and Safety at Work etc Act 1974, Sections 22–24 **Serial No.P**

PROHIBITION NOTICE

Name and address (See Section 46)	To ....
	....
(a) Delete as necessary	(a) Trading as ....
(b) Inspector's full name	(b) ....
(c) Inspector's official designation	one of (c)...
(d) Official address	of (d) ....
	.... Tel No. ....
	hereby give you notice that I am of the opinion that the following activities,
	namely:- ....
	....
	....
	where are (a) being carried on by you/about to be carried on by you/under your control
(e) Location of activity	at (e) ....
	involve, or will involve (a) a risk/an imminent risk, of serious personal injury. I am further of the opinion that the said matters involve contraventions of the following statutory provision:- ....
	....
	....
	....

because ....
....
....

and I hereby direct that the said activities shall not be carried on by you or under your control (a) immediately/after

(f) Date

(f)...

unless the said contraventions and matters included in the schedule, which forms part of this notice, have been remedied.

Signature .... Date.................

being an inspector appointed by an instrument in writing made pursuant to Section 19 of the said Act and entitled to issue this notice.

LP2

Appendix C
Protection of Eyes Regulations 1974

Schedule 1

SPECIFIED PROCESSES

Part 1

Processes in which approved eye protectors are required

All employers must provide eye protectors to all people employed in a factory on any of the following processes:

(i) the blasting or erosion of concrete by means of shot or other abrasive materials propelled by compressed air;

(ii) the cleaning of buildings or structures by means of shot or other abrasive materials propelled by compressed air;

(iii) cleaning by means of high-pressure water jets;

(iv) the striking of masonry nails by means of a hammer or other hand tool or by means of a power driven portable tool;

(v) any work carried out with a hand-held cartridge operated tool, including the operation of loading and unloading live cartridges into such a tool, and the handling of such a tool for the purpose of maintenance, repair or examination when the tool is loaded with a live cartridge;

(vi) the chipping of metal and the chipping, knocking out, cutting out or cutting off of cold rivets, bolts, nuts, lugs, pins, collars or similar articles from any structure or plant by means of a hammer, chisel, punch or similar hand tool, or by means of a power driven portable tool;

(vii) the chipping or scurfing of paint, scale, slag, rust or other corrosion from the surface of metal and other hard materials by means either of a hand tool or of a power driven portable tool, or by applying articles of metal or other such materials to a power driven tool;

(viii) the use of a high-speed metal cutting saw or an abrasive cutting-off wheel or disc, which in either case is power driven;

(ix) the pouring or skimming of molten metal in foundries;

(v) work at a molten salt bath when the molten salt surface is exposed;

(xi) the operation, maintenance or dismantling of any plant which contains acids, alkalis, dangerous corrosive substances, whether liquid or solid, or other substances which are similarly injurious to the eye and which have not been so prepared (by isolation, reduction of pressure, emptying or otherwise), treated or designed and constructed so as to prevent any reasonably foreseeable risk of injury to the eyes of any person engaged in any such work;

(xii) the handling in open vessels or manipulation of acids, alkalis, dangerous corrosive materials, whether liquid or solid, and other substances which are similarly injurious to the eyes, where there is a reasonably foreseeable risk of injury to the eyes of any person engaged in such work from drops splashed or particles thrown off;

(xiii) the driving in or on of bolts, pins, collars or similar articles to any structure or to any part of plant by means of a hammer, chisel, punch or similar hand tool or by means of a power driven portable tool, where there is a reasonably foreseeable risk of injury to the eyes of any person engaged in the work from particles or fragments thrown off;

(xiv) injection by pressure of liquids or solutions into buildings or structures where, in the course of such work, there is a reasonably foreseeable risk of injury to the eyes of any person engaged in the work from any such liquids or solutions;

(xv) the breaking up of metal by means of a hammer, whether power driven or not, or by means of a tup, where there is a reasonably foreseeable risk of injury to the eyes of any person engaged in the work from particles or fragments thrown off;

(xvi) the breaking, cutting, dressing, carving or drilling by means of a power-driven portable tool or by means of a hammer, chisel, pick or similar hand tool other than a trowel, of any of the following:

(1) glass, hard plastics, concrete, fired clay, plaster, slag or stone;

(2) materials similar to those in (1) above;

(3) articles consisting of those in (1) above;
(4) stonework, brickwork or blockwork;
(5) bricks, tiles or blocks (except blocks made of wood);

where in any of the above cases there is a reasonably foreseeable risk of injury to the eyes of any person engaged in the work from particles or fragments thrown off;

(xvii) the use of compressed air for removing swarf, dust, dirt or other particles, where, in the course of any such work, there is a reasonably foreseeable risk of injury to the eyes of any person engaged in the work from particles or fragments thrown off;

(xviii) work at a furnace containing molten metal, and the pouring or skimming of molten metal in places other than foundries, where there is a reasonably foreseeable risk of injury to the eyes of any person engaged in any such work from molten metal;

(xix) process in foundries where there is a reasonably foreseeable risk of injury to the eyes of any person engaged in any such work from hot sand thrown off;

(xx) work in the manufacture of wire and wire rope where there is a reasonably foreseeable risk of injury to the eyes of any person engaged in the work from particles or fragments thrown off or from flying ends of wire;

(xxi) the operation of coiling wire, and operations connected to it, where there is a reasonably foreseeable risk of injury to the eyes of any person engaged in any such work from particles or fragments thrown off or from flying ends of wire;

(xxii) the cutting of wire or metal strapping under tension, where there is a reasonably foreseeable risk of injury to the eyes of any person engaged in any such work from flying ends of wire or flying ends of metal strapping; and

(xxiii) work in the manufacture of glass and the processing of glass and the handling of cullet, where there is a reasonably foreseeable risk of injury to the eyes of any person engaged in the work from particles or fragments thrown off.

Part II
Processes in which approved shields or approved fixed shields are required

An employer must provide a sufficient number of fixed shields for employees in his factory engaged in any process involving the use of an exposed electric arc or an exposed stream of arc plasma.

Part III

Processes in which approved eye protectors or approved shields or approved fixed shields are required

The following are processes in which either approved eye protectors or approved shields or approved fixed shields are required:

(i) the welding of metals by means of apparatus to which oxygen or any flammable gas or vapour is supplied under pressure;

(ii) the hot fettling of steel castings by means of a flux-injected burner or air carbon torch, and the deseaming of metal;

(iii) the cutting, boring, cleaning, surface conditioning or spraying of material by means of apparatus (not being apparatus mechanically driven by compressed air) to which air, oxygen or any flammable gas or vapour is supplied under pressure excluding any such process elsewhere specified, where in any of these cases there is a reasonably foreseeable risk of injury to the eyes of any person engaged in the work from particles or fragments thrown off or from intense light or other radiation;

(iv) any process involving the use of an instrument which produces light amplification by the stimulated emission of radiation, being a process in which there is a reasonably foreseeable risk of injury to the eyes of any person engaged in the process from radiation;

(v) truing or dressing of an abrasive wheel where there is a reasonably foreseeable risk of injury to the eyes of any person engaged in the work from particles or fragments thrown off;

(vi) work with drop hammers, power hammers, horizontal forging machines and forging presses, other than hydraulic presses, used in any case for the manufacture of forgings;

(vii) the dry grinding of materials or articles by applying them by hand to a wheel, disc or band which in any such case is power driven or by means of a power-driven portable tool, where there is a reasonably foreseeable risk of injury to the eyes of any person engaged in the work from particles or fragments thrown off;

(viii) the fettling of metal castings involving the removal of metal, including runners, gates and risers, and the removal of any other material during the course of such fettling, where there is a reasonably foreseeable risk of injury to the eyes of any person engaged in the work from particles or fragments thrown off;

(ix) the production of metal castings at pressure die casting machines, where there is a reasonably foreseeable risk of injury to the eyes of any person engaged in any such work from molten metal thrown off;

(x) the machining of metals, including any dry grinding process not elsewhere specified, where there is a reasonably foreseeable risk of injury to the eyes of any person engaged in any work from particles or fragments thrown off; and

(xi) the welding of metals by an electric resistance process or a submerged electric arc, where there is a reasonably foreseeable risk of injury to the eyes of any person engaged in any such work from particles or fragments thrown off.

Schedule 2

Cases in which protection is required for persons at risk from, but who are not employed in the specified processes. Eye protectors or fixed shields must be proved

(i) the chipping of metal and the chipping, knocking out, cutting out or cutting off of cold rivets, bolts, nuts, lugs, pins, collars or similar articles from any structure or plant, by means of a hammer, chisel, punch or similar hand tool, or by means of a power-driven portable tool, where there is a reasonably foreseeable risk of injury to the eyes of any person not engaged in any such work from particles or fragments thrown off;

(ii) any process involving the use of an exposed electric arc or an exposed stream of arc plasma;

(iii) work with drop hammers, power hammers, horizontal forging machines and forging presses other than hydraulic presses used in any case for the manufacture of forgings;

(iv) the fettling of metal castings involving the removal of metal, including runners, gates and risers, and the removal of any other metal during the course of such fettling, where there is a reasonably foreseeable risk of injury to the eyes of any person not engaged in any such work from particles or fragments thrown off;

(v) any process involving the use of an instrument which produces light amplification by the stimulated emission of radiation (laser), where in any such process there is a reasonably foreseeable risk of injury to the eyes of the person not engaged in the process from radiation.

Appendix D
Prescribed industrial diseases and injuries

Diseases prescribed under the industrial injuries scheme and occupations for which they are prescribed

Description of disease or injury	Nature of occupation
Poisoning by:	*Any occupation involving:*
1. Lead or a compound of lead	The use or handling of, or exposure to the fumes, dust or vapour of, lead or a compound of lead, or a substance containing lead.
2. Manganese or a compound of manganese	The use or handling of, or exposure to the fumes, dust or vapour of, manganese or a compound of manganese, or a substance containing manganese.
3. Phosphorus or phosphine or poisoning due to the anti-cholinesterase action of organic phosphorus compounds	The use or handling of, or exposure to the fumes, dust or vapour of, phosphorus or a compound of phosphorus, or a substance containing phosporus.
4. Arsenic or a compound of arsenic	The use or handling of, or exposure to the fumes, dust or vapour of, arsenic or a compound of arsenic, or a substance containing arsenic.
5. Mercury or a compound of mercury	The use or handling of, or exposure to the fumes, dust or vapour of, mercury or a compound of mercury, or a substance containing mercury.

Description of disease or injury	Nature of occupation
Poisoning by:	*Any occupation involving:*
6. Carbon bisulphide	The use or handling of, or exposure to the fumes or vapour of, carbon bisulphide or a compound of carbon bisulphide, or a substance containing carbon bisulphide.
7. Benzene or a homologue	The use or handling of, or exposure to the fumes of, or vapour containing, benzene or any of its homologues.
8. A nitro- or amino- or chloro-derivative of benzene or of a homologue of benzene, or poisoning by nitrochlorbenzene	The use or handling of, or exposure to the fumes of, or vapour containing, a nitro- or amino- or chloro- derivative of benzene or a homologue of benzene or nitrochlorbenzene.
9. Dinitrophenol or a homologue or by substituted dinitrophenols or by the salts of such substances	The use or handling of, or exposure to the fumes of, or vapour containing, dinitrophenol or a homologue or substituted dinitrophenols or the salts of such substances.
10. Tetrachloroethane	The use or handling of, or exposure to the fumes of, or vapour containing, tetrachloroethane.
11. Tri-cresyl phosphate	The use or handling of, or exposure to the fumes of, or vapour containing, tri-cresyl phosphate.
12. Tri-phenyl phosphate	The use or handling of, or exposure to the fumes of, or vapour containing, tri-phenyl phosphate.
13. Diethylene dioxide (dioxan)	The use or handling of, or exposure to the fumes of, or vapour containing, diethylene dioxide (dioxan).
14. Methyl bromide	The use or handling of, or exposure to the fumes of, or vapour containing, methyl bromide.

Description of disease or injury	Nature of occupation
Poisoning by:	*Any occupation involving:*
15. Chlorinated naphthalene	The use or handling of, or exposure to the fumes of, or dust or vapour containing, chlorinated naphthalene.
16. Nickel carbonyl	Exposure to nickel carbonyl gas.
17. Nitrous fumes	The use or handling of nitric acid or exposure to nitrous fumes.
18. Gonioma kamassi (African box-wood)	The manipulation of gonioma kamassi or any process in or incidental to the manufacture of articles therefrom.
19. Anthrax	The handling of wool, hair, bristles, hides or skins or other animal products or residues, or contact with animals infected with anthrax.
20. Glanders	Contact with equine animals or their carcases.
21. Infection by all strains of leptospira	Work in place which are, or are liable to be, infested by rats, field mice or voles. Work involving contact with cattle and pigs (alive or dead) and dogs.
22. Ankylostomiasis	Work in or about a mine.
23. (a) Dystrophy of the corneal (including ulceration of the corneal surface) of the eye (b) Localised new growth of the skin, papillomatous or keratotic (c) Squamous-celled carcinoma of the skin *Due in any case to arsenic, tar, pitch, bitumen, mineral oil (including paraffin), soot or any compound, product (including quinone or hydroquinone), or residue of any of these substances*	The use or handling of, or exposure to, arsenic, tar, pitch, bitumen, mineral oil (including paraffin), soot or any compound, product (including quinone or hydroquinone), or residue of any of these substances.

Description of disease or injury	Nature of occupation
Poisoning by:	*Any occupation involving:*
25. Inflammation, ulceration or malignant disease of the skin or subcutaneous tissues or of the bones, or blood dyscrasia, or cataract, due to electro-magnetic radiations (other than radiant heat), or to ionising particles	Exposure to electro-magnetic radiations other than radiant heat, or to ionising particles.
26. Heat cataract	Frequent or prolonged exposure to rays from molten or red-hot material.
27. Decompression sickness	Subjection to compressed or rarefied air.
28. Cramp of the hand or forearm due to repetitive movements	Prolonged period of handwriting, typing or other repetitive movements of the fingers, hand or arm.
31. Subcutaneous cellulitis of the hand (Beat hand)	Manual labour causing severe or prolonged friction or pressure on the hand.
32. Bursitis or subcutaneous cellulitis arising at or about the knee due to severe or prolonged external friction or pressure at or about the knee (Beat knee)	Manual labour causing severe or prolonged external friction or pressure at or about the knee.
33. Bursitis or subcutaneous cellulitis arising at or about the elbow due to severe or prolonged external friction or pressure at or about the elbow (Beat elbow)	Manual labour causing severe or prolonged external friction or pressure at or about the elbow.
34. Traumatic inflammation of tendons of the hand or forearm or of the associated tendon sheaths	Manual labour, or frequent or repeated movements of the hand or wrist.
35. Miner's nystagmus	Work in or about a mine.
36. Poisoning by beryllium or a compound of beryllium	The use or handling of, or exposure to the fumes, dust or vapour of, beryllium or a compound of beryllium, or a substance containing beryllium.

Description of disease or injury	Nature of occupation
Poisoning by:	*Any occupation involving:*
37. (a) Carcinoma of the mucous membrane of the nose or associated air sinuses (b) Primary carcinoma of a bronchus or of a lung	Work in a factory where nickel is produced by decomposition of a gaseous nickel compound which necessitates working in or about a building or buildings where that process or any other industrial process ancillary or incidental thereto is carried on.
38. Tuberculosis	Close and frequent contact with a source or sources of tuberculosis infection by reason of employment: (a) in the medical treatment or nursing of a person or persons suffering from tuberculosis, or in a service ancillary to such treatment or nursing; (b) in attendance upon a person or persons suffering from tuberculosis, where the need for such attendance arises by reason of physical or mental infirmity; (c) as a research worker engaged in research in connection with tuberculosis; (d) as a laboratory worker, pathologist or person taking part in or assisting at post-mortem examinations of human remains where the occupation involves working with material which is a source of tuberculosis infection.
39. Primary neoplasm of the epithelial lining of the urinary bladder (Papilloma of the bladder), or of the renal pelvis or of the ureter or of the urethra	(a) Work in a building in which any of the following substances is produced for commercial purposes: (i) alpha-naphthylamine or beta-naphthylamine; (ii) diphenyl substituted by at least one nitro or primary amino group or by at least one nitro and primary amino group;

Description of disease or injury	Nature of occupation
Poisoning by:	*Any occupation involving:*
	(iii) any of the substances mentioned in sub-paragraph (ii) above if further ring substituted by halogeno, methyl or methoxy groups, but not by other groups; (iv) the salts of any of the substances mentioned in sub-paragraphs (i) to (iii) above; (v) auramine or magenta; (b) The use or handling of any of the substances mentioned in sub-paragraphs (i) to (iv) of paragraph (a), or work in a process in which any such substance is used or handled or is liberated; (c) the maintenance or cleaning of any plant or machinery used in any such process as is mentioned in paragraph (b), or the cleaning of clothing used in any such building as is mentioned in paragraph (a) if such clothing is cleaned within the works of which the building forms a part or in a laundry maintained and used solely in connection with such works.
40. Poisoning by cadmium	Exposure to cadmium fumes.
41. Inflammation or ulceration of the mucous membrane of the upper respiratory passages* or mouth produced by dust, liquid or vapour	Exposure to dust, liquid or vapour. **The upper respiratory passage are restricted to those of the nose, the pharynx and the larynx*
42. Non-infective dermatitis of external origin (including chrome ulceration of the skin but excluding dermatitis due to ionising particles or electro-magnetic radiations other than radiant heat)	Exposure to dust, liquid or vapour or any other external agent capable of irritating the skin (including friction or heat but excluding ionising particles or electro-magnetic radiations other than radiant heat).

Description of disease or injury	Nature of occupation
Poisoning by:	*Any occupation involving:*
43. Pulmonary disease due to the inhalation of the dust of mouldy hay or other mouldy vegetable produce and characterised by symptoms and signs attributable to a reaction in the peripheral part of the broncho-pulmonary system, and giving rise to a defect in gas exchange (Farmer's Lung)	Exposure to the dust of mouldy hay or other mouldy vegetable produce by reason of employment: (a) in agriculture, horticulture or forestry; or (b) loading or unloading or handling in storage such hay or other vegetable produce; or (c) handling bagasse.
44. Primary malignant neoplasm of the mesothelium (diffuse mesothelioma) of the pleura or of the peritoneum **Occupations involving asbestos may also give rise to asbestosis, a form of pneumoconiosis*	(a) The working or handling of asbestos* or any admixture of asbestos; (b) the manufacture or repair of asbestos textiles or other articles composed of asbestos; (c) the cleaning of any machinery or plant used in any of the foregoing operations and of any chambers, fixtures and appliances for the collection of asbestos dust; (d) substantial exposure to the dust arising from any of the foregoing operations.
45. Adeno-carcinoma of the nasal cavity or associated air sinuses	Attendance for work in or about a building where wooden furniture is manufactured.
46. Infection by brucella abortus	Contact with bovine animals infected by brucella abortus, their carcases or parts thereof or their untreated products, or with laboratory specimens or vaccines of or containing brucella abortus, by reason of employment: (a) as a farm worker; (b) as a veterinary worker; (c) as a slaughterhouse worker; (d) as a laboratory worker; or (e) in any work relating to the care, treatment, examination or handling of such animals, carcases or parts thereof, or products.

Description of disease or injury	**Nature of occupation**
Poisoning by:	*Any occupation involving:*
47. Poisoning by acrylamide monomer	The use or handling of, or exposure to, acrylamide monomer.
48. Substantial permanent sensorineural heating loss amounting to at least 50dB in each ear, being due in the case of at least one ear to occupational noise, and being the average of pure tone losses measured by audiometry over the 1, 2 and 3 kHz frequencies (occupational deafness)	(a) The use, or supervision of or assistance in the use, of pneumatic percussive tools, or the use of high-speed grinding tools, in the cleaning, dressing or finishing of cast metal or of ingots, billets of blooms; or (b) the use, or supervision of or assistance in the use, of pneumatic percussive tools on metal in the shipbuilding or ship repairing industries; or (c) the use, or supervision of or assistance in the use, of pneumatic percussive tools on metal, or for drilling rock in quarries or underground, or in coal-mining, for at least an average of one hour per working day; or (d) work wholly or mainly in the immediate vicinity of drop-forging plant (including plant for drop-stamping or drop-hammering) or forging press plant engaged in the shaping of hot metal; or (e) work wholly or mainly in rooms or sheds where there are machines engaged in weaving man-made or natural (including mineral) fibres, or in the bulking up of fibres in textile manufacture; or (f) the use of machines which cut, shape or clean metal nails; or (g) the use of plasma spray guns for the deposition of metal.
49. Viral hepatitis	(a) Close and frequent contact with human blood or human blood products; or

Description of disease or injury	Nature of occupation
Poisoning by:	*Any occupation involving:*
	(b) Close and frequent contact with a source of viral hepatitis infection by reason of employment in the medical treatment or nursing of a person suffering from viral hepatitis, or in a service ancillary to such treatment or nursing.
50. (a) Angiosarcoma of the liver 50. (b) Osteolysis of the terminal phalanges of the fingers	Work in or about machinery or apparatus for the polymerization of vinyl chloride monomer, a process which, for the purposes of this provision, comprises all operations up to and including the drying of the slurry produced by the polymerization and the packaging of the dried product; or work in a building or structure in which any part of the aforementioned process takes places.
51. Carcinoma of the nasal cavity or associated air sinuses (nasal carcinoma)	(a) attendance for work in a building used for the manufacture of footwear or components of footwear made wholly or partly of leather or fibre board; or (b) attendance for work at a place used wholly or mainly for the repair of footwear made wholly or partly of leather or fibre board.
52. Occupational vitiligo	Any occupation involving the use or handling of or exposure to para-tertiary-butylphenol, para-tertiary-butylcatechol, para-amylphenol, hydroquinane or the monobenzyl or monobutyl ether of hydroquinane.

NB Prescribed diseases nos 24, 29, and 30 have been deleted and combined with others.

Index

References are to paragraph numbers

References are to paragraph numbers

References are to paragraph numbers

References are to paragraph numbers

References are to paragraph numbers

References are to paragraph numbers